FUNDAMENTALS OF ENGINEERING NUMERICAL ANALYSIS

Engineers need hands-on experience in solving complex technical problems with computers. This text introduces numerical methods and shows how to develop, analyze, and use them. A thorough and practical book, it is intended as a first course in numerical analysis, primarily for beginning graduate students in engineering and physical science. Along with mastering the fundamentals of numerical methods, students will learn to write their own computer programs using standard numerical methods, or to use popular software packages such as MATLAB or programs in *Numerical Recipes* (Cambridge University Press). They will learn what factors affect accuracy, stability, and convergences, and they will also come to view critically the numerical output spewed from a computer. Numerous examples and exercises give students first-hand experience. The material is based on what Professor Moin found to be useful in teaching a course in numerical analysis and in his own career as a computational physicist/engineer.

With the availability of ever more powerful computers, numerical simulation of physical phenomena has become more practical and more widespread. This introductory text will guide students as well as practicing engineers in the use of computational teachings in their work.

Parviz Moin is Franklin P. and Caroline M. Johnson Professor of Engineering and Director of the Center for Turbulence Research, Stanford University. He pioneered the use of numerical simulation techniques for the study of turbulence physics, control, and modeling concepts. Professor Moin is a Fellow of the American Physical Society and a Member of the National Academy of Engineering.

FUNDAMENTALS OF ENGINEERING NUMERICAL ANALYSIS

PARVIZ MOIN

Stanford University

PUBLISHED BY THE PRESS SYNDICATE OF THE UNIVERSITY OF CAMBRIDGE
The Pitt Building, Trumpington Street, Cambridge, United Kingdom

CAMBRIDGE UNIVERSITY PRESS
The Edinburgh Building, Cambridge CB2 2RU, UK
40 West 20th Street, New York, NY 10011-4211, USA
10 Stamford Road, Oakleigh, VIC 3166, Australia
Ruiz de Alarcón 13, 28014 Madrid, Spain
Dock House, The Waterfront, Cape Town 8001, South Africa

http://www.cambridge.org

© Cambridge University Press 2001

This book is in copyright. Subject to statutory exception
and to the provisions of relevant collective licensing agreements,
no reproduction of any part may take place without
the written permission of Cambridge University Press.

First published 2001

Printed in the United States of America

Typefaces Times New Roman 10.75/13.5 pt., Melior and Helvetica Neue *System* LaTeX 2_ε [TB]

A catalog record for this book is available from the British Library.

Library of Congress Cataloging in Publication Data

Moin, Parviz,

Fundamentals of engineering numerical analysis / Parviz Moin.

p. cm.

ISBN 0-521-80140-0 – ISBN 0-521-80526-0 (pb.)

1. Engineering mathematics. 2. Numerical analysis. I. Title.

TA335 .M65 2001

620′.001′5194 – dc21 00-052933

ISBN 0 521 80140 0 hardback
ISBN 0 521 80526 0 paperback

To Linda

Contents

Preface

With the advent of faster computers, numerical simulation of physical phenomena is becoming more practical and more common. Computational prototyping is becoming a significant part of the design process for engineering systems. With ever-increasing computer performance the outlook is even brighter, and computer simulations are expected to replace expensive physical testing of design prototypes.

This book is an outgrowth of my lecture notes for a course in computational mathematics taught to first-year engineering graduate students at Stanford. The course is the third in a sequence of three quarter-courses in computational mathematics. The students are expected to have completed the first two courses in the sequence: numerical linear algebra and elementary partial differential equations. Although familiarity with linear algebra in some depth is essential, mastery of the analytical tools for the solution of partial differential equations (PDEs) is not; only familiarity with PDEs as governing equations for physical systems is desirable. There is a long tradition at Stanford of emphasizing that engineering students learn numerical analysis (as opposed to learning to run canned computer codes). I believe it is important for students to be educated about the fundamentals of numerical methods. My first lesson in numerics includes a warning to the students not to believe, at first glance, the numerical output spewed out from a computer. They should know what factors affect accuracy, stability, and convergence and be able to ask tough questions before accepting the numerical output. In other words, the user of numerical methods should not leave all the "thinking" to the computer program and the person who wrote it. It is also important for computational physicists and engineers to have first-hand experience with solving real problems with the computer. They should experience both the power of numerical methods for solving non-trivial problems as well as the frustration of using inadequate methods. Frustrating experiences with a numerical method almost always send a competent numerical analyst to the drawing board and force him or her to ask good questions about the choice and parameters of the method, which should have been asked before going to the computer in the first place. The exercises at the end of each chapter are intended to give these important experiences with numerical methods.

Along with mastering the fundamentals of numerical methods, the students are expected to write their own programs to solve problems using standard numerical methods. They are also encouraged to use standard (commercial) software whenever possible. There are several software libraries with well-documented programs for basic computational work. Recently, I have used the *Numerical Recipes* by Press et al. (Cambridge) as an optional supplement to my lectures. *Numerical Recipes* is based on a large software library that is well documented and available on computer disks. Some of the examples in this book refer to specific programs in *Numerical Recipes*.

Students should also have a simple (x, y) plotting package to display their numerical results. Some students prefer to use MATLAB's plotting software, some use the plotting capability included with a spreadsheet package, and others use more sophisticated commercial plotting packages. Standard well-written numerical analysis programs are generally available for almost everything covered in the first four chapters, but this is not the case for partial differential equations discussed in Chapter 5. The main technical reason for this is the large variety of partial differential equations, which requires essentially tailor-made programs for each application.

No attempt has been made to provide complete coverage of the topics that I have chosen to include in this book. This is not meant to be a reference book, rather it contains the material for a first course in numerical analysis for future practitioners. Most of the material is what I have found useful in my career as a computational physicist/engineer. The coverage is succinct, and it is expected that all the material will be covered sequentially. The book is intended for first-year graduate students in science and engineering or seniors with good post-calculus mathematics backgrounds. The first five chapters can be covered in a one-quarter course, and Chapter 6 can be included in a one-semester course.

Discrete data and numerical interpolation are introduced in Chapter 1, which exposes the reader to the dangers of high-order polynomial interpolation. Cubic splines are offered as a good working algorithm for interpolation. Chapter 2 (finite differences) and Chapter 3 (numerical integration) are the foundations of discrete calculus. Here, I emphasize systematic procedures for constructing finite difference schemes, including high-order Padé approximations. We also examine alternative, and often more informative, measures of numerical accuracy. In addition to introducing the standard numerical integration techniques and their error analysis, we show in Chapter 3 how knowledge of the form of numerical errors can be used to construct more accurate numerical results (Richardson extrapolation) and to construct adaptive schemes that obtain the solution to the accuracy specified by the user. Usually, at this point in my lectures, I seize the opportunity, offered by these examples, to stress the value of a detailed knowledge of numerical error and its pay-offs even for the most application-oriented students. Knowledge is quickly transferred to power in constructing novel numerical methods.

Chapter 4 is on numerical solution of ordinary differential equations (ODEs) – the heart of this first course in numerical analysis. A number of new concepts such as stability and stiffness are introduced. The reader begins to experience new tools in his arsenal for solving relatively complex problems that would have been impossible to do analytically. Because so many interesting applications are cast in ordinary differential equations, this chapter is particularly interesting for engineers. Different classes of numerical methods are introduced and analyzed even though there are several well-known powerful numerical ODE solver packages available to solve any practical ODE without having to know their inner workings. The reason for this extensive coverage of a virtually solved problems is that the same algorithms are used for solution of partial differential equations when canned programs for general PDEs are not available and the user is forced to write his or her own programs. Thus, it is essential to learn about the properties of numerical methods for ODEs in order to develop good programs for PDEs.

Chapter 5 discusses the numerical solution of partial differential equations and relies heavily on the analysis of initial value problems introduced for ODEs. In fact by using the modified wavenumber analysis, we can cast into ODEs the discretized initial value problems in PDEs, and the knowledge of ODE properties becomes very useful and no longer of just academic value. Once again the knowledge of numerical errors is used to solve a difficult problem of dealing with large matrices in multi-dimensional PDEs by the approximate factorization technique. Dealing with large matrices is also a focus of numerical techniques for elliptic partial differential equations, which are dealt with by introducing the foundations of iterative solvers.

Demand for high accuracy is increasing as computational engineering matures. Today's engineers and physicists are less interested in qualitative features of numerical solutions and more concerned with numerical accuracy. A branch of numerical analysis deals with spectral methods, which offer highly accurate numerical methods for solution of partial differential equations. Chapter 6 covers aspects of Fourier analysis and introduces transform methods for partial differential equations.

My early work in numerical analysis was influenced greatly by discussions with Joel Ferziger and subsequently by the works of Harvard Lomax at NASA–Ames. Thanks are due to all my teaching assistants who helped me develop the course upon which this book is based; in particular, I thank Jon Freund and Arthur Kravchenko who provided valuable assistance in preparation of this book. I am especially grateful to Albert Honein for his substantial help in preparing this book in its final form and for his many contributions as my teaching assistant in several courses in computational mathematics at Stanford.

Parviz Moin
Stanford, California
July 2000

1

Interpolation

Often we want to fit a smooth curve through a set of data points. Applications might be differentiation or integration or simply estimating the value of the function between two adjacent data points. With interpolation we actually pass a curve *through* the data. If data are from a crude experiment characterized by some uncertainty, it is best to use the method of least squares, which does not require the approximating function to pass through all the data points.

1.1 Lagrange Polynomial Interpolation

Suppose we have a set of $n + 1$ (not necessarily equally spaced) data (x_i, y_i). We can construct a polynomial of degree n that passes through the data:

$$P(x) = a_0 + a_1 x + a_2 x^2 + \cdots + a_n x^n.$$

The $n + 1$ coefficients of P are determined by forcing P to pass through the data. This leads to $n + 1$ equations in the $n + 1$ unknowns, $a_0, a_1, \ldots, a_n$:

$$y_i = P(x_i) = a_0 + a_1 x_i + a_2 x_i^2 + \cdots + a_n x_i^n \quad i = 0, 1, 2, \ldots, n.$$

This procedure for finding the coefficients of the polynomial is not very attractive. It involves solving a system of algebraic equations that is generally ill-conditioned (see Appendix) for large n. In practice we will define the polynomial in an explicit way (as opposed to solving a system of equations). Consider the following polynomial of degree n associated with each point x_j:

$$L_j(x) = \alpha_j (x - x_0)(x - x_1) \cdots (x - x_{j-1})(x - x_{j+1}) \cdots (x - x_n),$$

where α_j is a constant to be determined. In the product notation, L_j is written as follows

$$L_j(x) = \alpha_j \prod_{\substack{i=0 \\ i \neq j}}^{n} (x - x_i).$$

If x is equal to any of the data points except x_j, then $L_j(x_i) = 0$ for $i \neq j$. For $x = x_j$,

$$L_j(x_j) = \alpha_j \prod_{\substack{i=0 \\ i \neq j}}^{n} (x_j - x_i).$$

We now define α_j to be

$$\alpha_j = \left[\prod_{\substack{i=0 \\ i \neq j}}^{n} (x_j - x_i) \right]^{-1}.$$

Then, L_j will have the following important property:

$$L_j(x_i) = \begin{cases} 0 & x_i \neq x_j \\ 1 & x_i = x_j. \end{cases} \tag{1.1}$$

Next we form a linear combination of these polynomials with the data as weights:

$$P(x) = \sum_{j=0}^{n} y_j L_j(x). \tag{1.2}$$

This is a polynomial of degree n because it is a linear combination of polynomials of degree n. It is called a *Lagrange polynomial*. It is the desired interpolating polynomial because by construction, it passes through all the data points. For example, at $x = x_i$

$$P(x_i) = y_0 L_0(x_i) + y_1 L_1(x_i) + \cdots + y_i L_i(x_i) + \cdots + y_n L_n(x_i).$$

Since $L_i(x_k)$ is equal to zero except for $k = i$, and $L_i(x_i) = 1$,

$$P(x_i) = y_i.$$

Note that polynomial interpolation is unique. That is, there is only one polynomial of degree n that passes through a set of $n + 1$ points*. The Lagrange polynomial is just a compact, numerically better behaved way of expressing the polynomial whose coefficients could have also been obtained from solving a system of algebraic equations.

For a large set of data points (say greater than 10), polynomial interpolation for uniformly spaced data can be very dangerous. Although the polynomial is fixed (tied down) at the data points, it can wander wildly between them, which can lead to large errors for derivatives or interpolated values.

* The uniqueness argument goes like this: suppose there are two polynomials of degree n, Z_1 and Z_2 that pass through the same data points, $x_0, x_1, \ldots, x_n$. Let $Z = Z_1 - Z_2$. Z is a polynomial of degree n with $n + 1$ zeros, $x_0, x_1, \ldots, x_n$, which is impossible unless Z is identically zero.

EXAMPLE 1.1 Lagrange Interpolation

Consider the following data, which are obtained from a smooth function also known as Runge's function, $y = (1 + 25x^2)^{-1}$:

x_i	−1.00	−0.80	−0.60	−0.40	−0.20	0.00	0.20	0.40	0.60	0.80	1.00
y_i	0.038	0.058	0.100	0.200	0.500	1.00	0.500	0.200	0.100	0.058	0.038

We wish to fit a smooth curve through the data using the Lagrange polynomial interpolation, for which the value at any point x is simply

$$P(x) = \sum_{j=0}^{n} y_j \prod_{\substack{i=0 \\ i \neq j}}^{n} \frac{x - x_i}{x_j - x_i}.$$

For example at the point $(x = 0.7)$, the interpolated value is

$$\begin{aligned} P(.7) = {} & 0.038\frac{(0.7+0.8)(0.7+0.6)\cdots(0.7-0.8)(0.7-1.0)}{(-1.0+0.8)(-1.0+0.6)\cdots(-1.0-0.8)(-1.0-1.0)} \\ & +0.058\frac{(0.7+1.0)(0.7+0.6)\cdots(0.7-0.8)(0.7-1.0)}{(-0.8+1.0)(-0.8+0.6)\cdots(-0.8-0.8)(-0.8-1.0)} \\ & +\cdots \\ & +0.038\frac{(0.7+1.0)(0.7+0.8)\cdots(0.7-0.6)(0.7-0.8)}{(1.0+1.0)(1.0+0.6)\cdots(1.0-0.6)(1.0-0.8)} = -0.226. \end{aligned}$$

Evaluating the interpolating polynomial at a large number of intermediate points, we may plot the resulting polynomial curve passing through the data points (see Figure 1.1). It is clear that the Lagrange polynomial behaves very poorly between some of the data points, especially near the ends of the interval. The problem does not go away by simply having more data points in the interval and thereby tying down the function further. For example, if instead of eleven points we had twenty-one uniformly spaced data points in the same interval, the overshoots at the ends would have peaked at nearly 60 rather than at 1.9 as they did for eleven points. However, as shown in the following example, the problem can be somewhat alleviated if the data points are non-uniformly spaced with finer spacings near the ends of the interval.

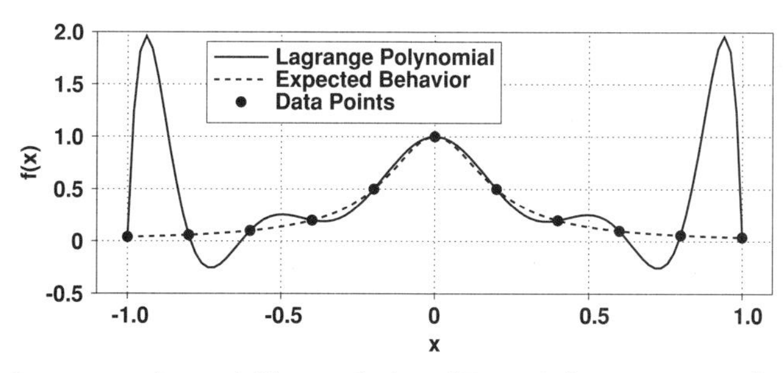

Figure 1.1 Lagrange polynomial interpolation of Runge's function using eleven equally spaced data points.

EXAMPLE 1.2 Lagrange Interpolation With Non-equally Spaced Data

Consider the following data which are again extracted from the Runge's function of Example 1. The same number of points are used as in Example 1.1, but the data points x_i are now more finely spaced near the ends (at the expense of coarser resolution near the center).

x_i	−1.00	−0.95	−0.81	−0.59	−0.31	0.00	0.31	0.59	0.81	0.95	1.00
y_i	0.038	0.042	0.058	0.104	0.295	1.00	0.295	0.104	0.058	0.042	0.038

The interpolation polynomial and the expected curve, which in this case (as in Example 1) is simply the Runge's function, are plotted in Figure 1.2. It is apparent that the magnitudes of the overshoots at the ends of the interval have been reduced; however, the overall accuracy of the scheme is still unacceptable.

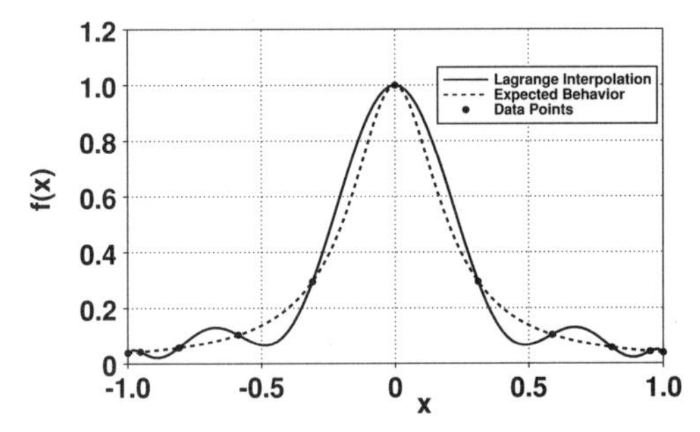

Figure 1.2 Lagrange polynomial interpolation of Runge's function using eleven non-equally spaced data points. The data toward the ends of the interval are more finely spaced.

The wandering problem can also be severely curtailed by *piecewise Lagrange* interpolation. Instead of fitting a single polynomial of degree n to all the data, one fits lower order polynomials to sections of it. This is used in many practical applications and is the basis for some numerical methods. The main problem with piecewise Lagrange interpolation is that it has discontinuous slopes at the boundaries of the segments, which causes difficulties when evaluating the derivatives at the data points. Interpolation with cubic splines circumvents this difficulty.

1.2 Cubic Spline Interpolation

Interpolation with cubic splines is essentially equivalent to passing a flexible plastic ruler through the data points. You can actually hammer a few nails partially into a board and pretend that they are a set of data points; the nails can then hold a plastic ruler that is bent to touch all the nails. Between the nails, the ruler acts as the interpolating function. From mechanics the equation governing

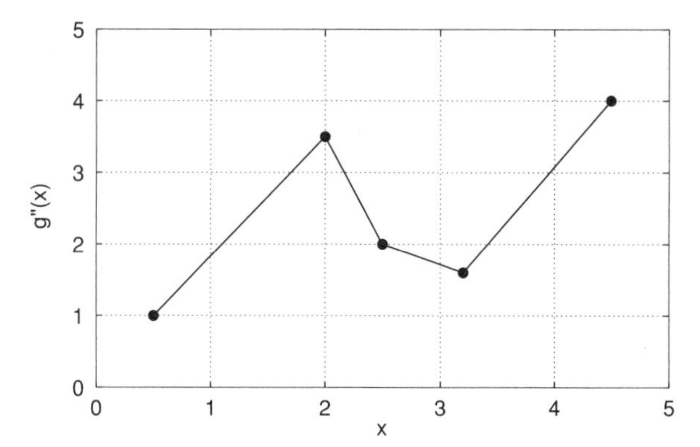

Figure 1.3 A schematic showing the linearity of g'' in between the data points. Also note that with such a construction, g'' is continuous at the data points.

the position of the curve $y(x)$ traced by the ruler is

$$Cy^{(iv)} = G,$$

where C depends on the material properties and G represents the applied force necessary to pass the spline through the data. The force is applied only at the data points; between the data points the force is zero. Therefore, the spline is piecewise cubic between the data. As will be shown below, the spline interpolant and its *first two derivatives are continuous at the data points.*

Let $g_i(x)$ be the cubic in the interval $x_i \leq x \leq x_{i+1}$ and let $g(x)$ denote the collection of all the cubics for the entire range of x. Since g is piecewise cubic its second derivative, g'', is piecewise linear. For the interval $x_i \leq x \leq x_{i+1}$, we can write the equation for the corresponding straight line as

$$g_i''(x) = g''(x_i)\frac{x - x_{i+1}}{x_i - x_{i+1}} + g''(x_{i+1})\frac{x - x_i}{x_{i+1} - x_i}. \tag{1.3}$$

Note that by construction, in (1.3) we have enforced the continuity of the second derivative at the data points. That is, as shown in Figure 1.3, straight lines from the adjoining intervals meet at the data points.

Integrating (1.3) twice we obtain

$$g_i'(x) = \frac{g''(x_i)}{x_i - x_{i+1}}\frac{(x - x_{i+1})^2}{2} + \frac{g''(x_{i+1})}{x_{i+1} - x_i}\frac{(x - x_i)^2}{2} + C_1 \tag{1.4}$$

and

$$g_i(x) = \frac{g''(x_i)}{x_i - x_{i+1}}\frac{(x - x_{i+1})^3}{6} + \frac{g''(x_{i+1})}{x_{i+1} - x_i}\frac{(x - x_i)^3}{6} + C_1 x + C_2. \tag{1.5}$$

The undetermined constants C_1 and C_2 are obtained by matching the functional values at the end points:

$$g_i(x_i) = f(x_i) \qquad g_i(x_{i+1}) = f(x_{i+1}),$$

which give two equations for the two unknowns, C_1 and C_2. Substituting for C_1

and C_2 in (1.5) leads to the spline equation used for interpolation:

$$g_i(x) = \frac{g''(x_i)}{6}\left[\frac{(x_{i+1}-x)^3}{\Delta_i} - \Delta_i(x_{i+1}-x)\right] + \frac{g''(x_{i+1})}{6}\left[\frac{(x-x_i)^3}{\Delta_i} - \Delta_i(x-x_i)\right] + f(x_i)\frac{x_{i+1}-x}{\Delta_i} + f(x_{i+1})\frac{x-x_i}{\Delta_i}, \tag{1.6}$$

where $x_i \le x \le x_{i+1}$ and $\Delta_i = x_{i+1} - x_i$. In (1.6) $g''(x_i)$ and $g''(x_{i+1})$ are still unknowns. To obtain $g''(x_i)$, we use the remaining matching condition, which is the continuity of the first derivatives:

$$g_i'(x_i) = g_{i-1}'(x_i).$$

The desired system of equations for $g''(x_i)$ is then obtained from (1.4) by evaluating it at $x = x_i$ and equating the result to the expression obtained by replacing i with $i - 1$. This leads to

$$\frac{\Delta_{i-1}}{6}g''(x_{i-1}) + \frac{\Delta_{i-1}+\Delta_i}{3}g''(x_i) + \frac{\Delta_i}{6}g''(x_{i+1}) = \frac{f(x_{i+1})-f(x_i)}{\Delta_i} - \frac{f(x_i)-f(x_{i-1})}{\Delta_{i-1}} \qquad i = 1, 2, 3, \ldots, N-1. \tag{1.7}$$

These are $N - 1$ equations for the $N + 1$ unknowns $g''(x_0),\ g''(x_1), \ldots,\ g''(x_N)$. The equations are in tridiagonal form and diagonally dominant, and therefore they can be solved very efficiently. The remaining equations are obtained from the prescription of some "end conditions." Typical conditions are:

a) Free run-out (natural spline):

$$g''(x_0) = g''(x_N) = 0.$$

This is the most commonly used condition. It can be shown that with this condition, the spline is the smoothest interpolant in the sense that the integral of g''^2 over the whole interval is smaller than any other function interpolating the data.

b) Parabolic run-out:

$$g''(x_0) = g''(x_1)$$
$$g''(x_{N-1}) = g''(x_N).$$

In this case, the interpolating polynomials in the first and last intervals are parabolas rather than cubics (see Exercise 3).

c) Combination of (a) and (b):

$$g''(x_0) = \alpha g''(x_1)$$
$$g''(x_{N-1}) = \beta g''(x_N),$$

where α and β are constants chosen by the user.

d) Periodic:

$$g''(x_0) = g''(x_{N-1})$$
$$g''(x_1) = g''(x_N).$$

This condition is suitable for interpolating in one period of a known periodic signal.

The general procedure for spline interpolation is first to solve the system of equations (1.7) with the appropriate end conditions for $g''(x_i)$. The result is then used in (1.6), providing the interpolating function $g_i(x)$ for the interval $x_i \leq x \leq x_{i+1}$. In general, spline interpolation is preferred over Lagrange polynomial interpolation; it is easy to implement and usually leads to smooth curves.

EXAMPLE 1.3 Cubic Spline Interpolation

We will now interpolate the data in Example 1 with a cubic spline. We solve the tridiagonal system derived in (1.7). Since the data are uniformly spaced, this equation takes a particularly simple form for $g''(x_i)$:

$$\frac{1}{6}g''(x_{i-1}) + \frac{2}{3}g''(x_i) + \frac{1}{6}g''(x_{i+1}) = \frac{y_{i+1} - 2y_i + y_{i-1}}{\Delta^2} \qquad i = 1, 2, \ldots, n-1.$$

For this example, we will use the free run-out condition $g''(x_0) = g''(x_n) = 0$. The cubic spline is evaluated at several x points using (1.6) and the $g''(x_i)$ values obtained from the solution of this tridiagonal system. The subroutine `spline` in *Numerical Recipes* has been used in the calculation. The result is presented in Figure 1.4. Spline representation appears to be very smooth and is virtually indistinguishable from Runge's function.

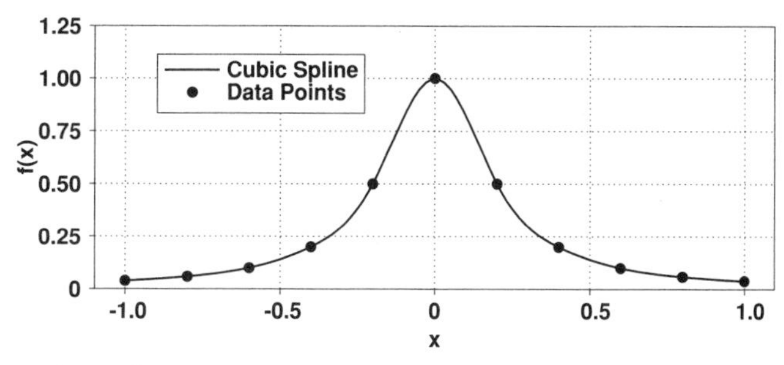

Figure 1.4 Cubic spline interpolation of Runge's function using the equally spaced data of Example 1.

Clearly spline interpolation is much more accurate than Lagrange interpolation. Of course, the computer program for spline is longer and a bit more complicated than that for Lagrange interpolation. However, once such programs are written for general use, then the time taken to develop the program, or the "human cost," no longer enters into consideration.

An interesting version of spline interpolation, called tension spline, can be used if the spline fit wiggles too much. The idea is to apply some tension or pull from both ends of the flexible ruler discussed at the beginning of this section. Mathematically, this also leads to a tridiagonal system of equations for g_i'', but the coefficients are more complicated. In the limit of very large tension, all the wiggles are removed, but the spline is reduced to a simple straight line interpolation (see Exercise 6).

EXERCISES

1. Write a computer program for Lagrange interpolation (you may want to use the *Numerical Recipes* subroutine `polint`). Test your program by verifying that $P(0.7) = -0.226$ in Example 1.
   (a) Using the data of Example 1, find the interpolated value at $x = 0.9$.
   (b) Use Runge's function to generate a table of 21 equally spaced data points. Interpolate these data using a Lagrange polynomial of order 20. Plot this polynomial and comment on the comparison between your result and the plot of Example 1.
2. Derive an expression for the derivative of a Lagrange polynomial of order n at a point x between the data points.
3. Show that if parabolic run-out conditions are used for cubic spline interpolation, then the interpolating polynomials in the first and last intervals are indeed parabolas.
4. An operationally simpler spline is the so-called quadratic spline. Interpolation is carried out by piecewise quadratics.
   (a) What are the suitable joint conditions for quadratic spline?
   (b) Show how the coefficients of the spline are obtained. What are suitable end conditions?
   (c) Compare the required computational efforts for quadratic and cubic splines.
5. Consider a set of $n + 1$ data points $(x_0, f_0), \ldots, (x_n, f_n)$, equally spaced with $x_{i+1} - x_i = h$. Discuss how cubic splines can be used to obtain a numerical approximation for the first derivative f' at these data points. Give a detailed account of the required steps. You should obtain formulas for the numerical derivative at the data points $x_0, \ldots, x_n$ and explain how to calculate the terms in the formulas.
6. Tension splines can be used if the interpolating spline wiggles too much. In this case, the equation governing the position of the plastic ruler in between the data points is

$$y^{(iv)} - \sigma^2 y'' = 0$$

   where σ is the tension parameter. If we denote $g_i(x)$ as the interpolating tension spline in the interval $x_i \leq x \leq x_{i+1}$, then $g_i''(x) - \sigma^2 g_i(x)$ is a straight line in this interval, which can be written in the following convenient forms:

$$g_i''(x) - \sigma^2 g_i(x) = [g''(x_i) - \sigma^2 f(x_i)]\frac{x - x_{i+1}}{x_i - x_{i+1}} + [g''(x_{i+1}) - \sigma^2 f(x_i + 1)]\frac{x - x_i}{x_{i+1} - x_i}.$$

(a) Verify that for $\sigma = 0$, the cubic spline is recovered, and $\sigma \to \infty$ leads to linear interpolation.

(b) Derive the equation for tension spline interpolation, i.e., the expression for $g_i(x)$.

7. The tuition for nine units at *St. Anford* University has been increasing from 1980 to 1998 as shown in the table below:

Year	**Tuition per year**
1980	\$3,516
1982	\$4,541
1984	\$5,483
1986	\$6,550
1988	\$7,825
1990	\$9,354
1992	\$10,816
1994	\$12,180
1996	\$13,509
1998	\$14,911

(a) Plot the given data points and intuitively interpolate (draw) a smooth curve through them.

(b) Interpolate the data with the Lagrange polynomial. Plot the polynomial and the data points. Use the polynomial to predict the tuition in 2001. This is an extrapolation problem; discuss the utility of Lagrange polynomials for extrapolation.

(c) Repeat (b) with a cubic spline interpolation and compare your results.

8. The concentration of a certain toxin in a system of lakes downwind of an industrial area has been monitored very accurately at intervals from 1978 to 1992 as shown in the table below. It is believed that the concentration has varied smoothly between these data points.

Year	**Toxin Concentration**
1978	12.0
1980	12.7
1982	13.0
1984	15.2
1986	18.2
1988	19.8
1990	24.1
1992	28.1
1994	???

(a) Interpolate the data with the Lagrange polynomial. Plot the polynomial and the data points. Use the polynomial to predict the condition of the lakes in 1994. Discuss this prediction.
(b) Interpolation may also be used to fill "holes" in the data. Say the data from 1982 and 1984 disappeared. Predict these values using the Lagrange polynomial fitted through the other known data points.
(c) Repeat (b) with a cubic spline interpolation. Compare and discuss your results.

9. Consider a piecewise Lagrange polynomial that interpolates between three points at a time. Let a typical set of consecutive three points be x_{i-1}, x_i, and x_{i+1}. Derive differentiation formulas for the first and second derivatives at x_i. Simplify these expressions for uniformly spaced data with $\Delta = x_{i+1} - x_i$. You have just derived finite difference formulas for discrete data, which are discussed in the next chapter.
10. Consider a function f defined on a set of $N + 1$ discrete points

$$x_0 < x_1 < \cdots < x_N.$$

We want to derive an $(N + 1) \times (N + 1)$ matrix, D (with elements d_{ij}), which when multiplied by the vector of the values of f on the grid results in the derivative of f' at the grid points. Consider the Lagrange polynomial interpolation of f in (1.2):

$$P(x) = \sum_{j=0}^{N} y_j L_j(x).$$

We can differentiate this expression to obtain P'. We seek a matrix D such that

$$D\boldsymbol{f} = \boldsymbol{P}'_N$$

where, $\boldsymbol{P}'_N$ is a vector whose elements are the derivative of $\boldsymbol{P}_N$ at the data points. Note that the derivative approximation given by $D\boldsymbol{f}$ is exact for all polynomials of degree N or less. We define D such that it gives the exact derivatives for all such polynomials at the $N + 1$ grid points. That is, we want

$$D \underbrace{L_k(x_j)}_{\delta_{kj}} = L'_k(x_j) \qquad j, k = 0, 1, 2, \ldots, N$$

where δ_{kj} is Kronecker delta which is equal to one for $k = j$ and zero for $k \neq j$. Show that this implies that

$$d_{jk} = \left. \frac{d}{dx} L_k \right|_{x=x_j}, \tag{1}$$

where d_{jk} are the elements of D. Evaluate the right-hand side of (1) and show that

$$d_{jk} = L'_k(x_j) = \alpha_k \prod_{\substack{l=0 \\ l \neq j,k}}^{N} (x_j - x_l) = \frac{\alpha_k}{\alpha_j (x_j - x_k)} \qquad \text{for } j \neq k, \tag{2}$$

and

$$d_{jj} = L'_j(x_j) = \sum_{\substack{l=0 \\ l \neq j}}^{N} \frac{1}{x_j - x_l} \qquad \text{for } j = k \tag{3}$$

where, α_j is defined in Section 1.1.
(HINT: Take the logarithm of $L_k(x)$.)

FURTHER READING

Dahlquist, G., and Björck, Å. *Numerical Methods*. Prentice-Hall, 1974, Chapters 4 and 7.

Ferziger, J. H. *Numerical Methods for Engineering Application*, Second Edition. Wiley, 1998, Chapter 2.

Forsythe, G. E., Malcolm, M. A., and Moler, C. B. *Computer Methods for Mathematical Computations*. Prentice-Hall, 1977, Chapter 4.

Press, W. H., Teukolsky, S. A., Vetterling, W. T., and Flannery, B. P. *Numerical Recipes: The Art of Scientific Computing*, Second Edition. Cambridge University Press, 1992, Chapter 3.

2

Numerical Differentiation – Finite Differences

In the next two chapters we develop a set of tools for discrete calculus. This chapter deals with the technique of finite differences for numerical differentiation of discrete data. We develop and discuss formulas for calculating the derivative of a smooth function, but only as defined on a discrete set of grid points $x_0, x_1, \ldots, x_N$. The data may already be tabulated or a table may have been generated from a complicated function or a process. We will focus on finite difference techniques for obtaining numerical values of the derivatives at the grid points. In Chapter 6 another more elaborate technique for numerical differentiation is introduced. Since we have learned from calculus how to differentiate any function, no matter how complicated, finite differences are seldom used for approximating the derivatives of explicit functions. This is in contrast to integration, where we frequently have to look up integrals in tables, and often solutions are not known. As will be seen in Chapters 4 and 5, the main application of finite differences is for obtaining numerical solution of differential equations.

2.1 Construction of Difference Formulas Using Taylor Series

Finite difference formulas can be easily derived from Taylor series expansions. Let us begin with the simplest approximation for the derivative of $f(x)$ at the point x_j, we use the Taylor series:

$$f(x_{j+1}) = f(x_j) + (x_{j+1} - x_j)f'(x_j) + \frac{(x_{j+1} - x_j)^2}{2} f''(x_j) + \cdots. \quad (2.1)$$

Rearrangement leads to

$$f'(x_j) = \frac{f(x_{j+1}) - f(x_j)}{\Delta x_j} - \frac{\Delta x_j}{2} f''(x_j) + \cdots \quad (2.2)$$

where $\Delta x_j = x_{j+1} - x_j$ is the mesh size. The first term on the right-hand side of (2.2) is a finite difference approximation to the derivative. The next term is the *leading error* term. In this book, we also use h to indicate the mesh size. When the grid points are uniformly spaced, no subscript will be attached to h or Δx.

Formula (2.2) is usually recast in the following form

$$f'_j = \frac{f_{j+1} - f_j}{h} + O(h), \tag{2.3}$$

which is referred to as the first-order *forward difference*. This is the same expression used to define the derivative in calculus, except that in calculus the definition involves the limit, $h \to 0$; but here, h is finite.

The exponent of h in $O(h^\alpha)$ is the order of accuracy of the method. It is a useful measure of accuracy because it gives an indication of how rapidly the accuracy can be improved with refinement of the grid spacing. For example, with a first-order scheme, such as in (2.3), if we reduce the mesh size by a factor of 2, the error (called the *truncation error*) is reduced by approximately a factor of 2. Notice that when we talk about the truncation error of a finite difference scheme, we always refer to the leading error term with the implication that the higher order terms in the Taylor series expansion are much smaller than the leading term. That is, for sufficiently small h the higher powers of h, which appear as coefficients of the other terms, get smaller. Of course, one should not be concerned with the actual value of h in dimensional units; for example, h can be in tens of kilometers in atmospheric dynamics problems, which may lead to the concern that the higher order terms that involve higher powers of h become larger. This apparent dilemma is quickly overcome by non-dimensionalizing the dependent variable x in (2.1). Let us non-dimensionalize x with the domain length $L = x_N - x_0$. L is actually cancelled out in the non-dimensionalization of (2.1), but now we would be certain that the non-dimensional increment $x_{j+1} - x_j$ is always less than 1, and hence, its higher powers get smaller.

Let us now consider some other popular finite difference formulas. By expanding f_{j-1} about x_j, we can get

$$f'_j = \frac{f_j - f_{j-1}}{h} + O(h), \tag{2.4}$$

which is also a first-order scheme, called the first-order *backward difference* formula. Higher order schemes (more accurate) can be derived by Taylor series of the function f at different points about the point x_j. For example, the widely used *central difference* formula can be obtained from subtraction of two Taylor series expansions; assuming uniformly spaced data we have

$$f_{j+1} = f_j + hf'_j + \frac{h^2}{2}f''_j + \frac{h^3}{6}f'''_j + \cdots \tag{2.5}$$

$$f_{j-1} = f_j - hf'_j + \frac{h^2}{2}f''_j - \frac{h^3}{6}f'''_j + \cdots, \tag{2.6}$$

which leads to

$$f'_j = \frac{f_{j+1} - f_{j-1}}{2h} - \frac{h^2}{6} f'''_j + \cdots. \tag{2.7}$$

This is, of course, a second-order formula. That is, if we refine the mesh by a factor of 2, we expect the truncation error to reduce by a factor of 4. In general, we can obtain higher accuracy if we include more points. Here is a fourth-order formula:

$$f'_j = \frac{f_{j-2} - 8f_{j-1} + 8f_{j+1} - f_{j+2}}{12h} + O(h^4). \tag{2.8}$$

The main difficulty with higher order formulas occurs near boundaries of the domain. They require the functional values at points outside the domain, which are not available. For example, if the values of the function f are known at points $x_0, x_1, \ldots, x_N$ and the derivative of f at x_1 is required, the formula (2.8) would require the value of f at x_{-1} (in addition to x_0, x_1, x_2, and x_3) which is not available. In practice, to alleviate this problem, we utilize lower order or non-central formulas near boundaries. Similar formulas can be derived for second- or higher order derivatives. For example, the second-order central difference formula for the second derivative is derived by adding (2.5) and (2.6), the two f'_j terms are cancelled, and after a minor rearrangement, we get

$$f''_j = \frac{f_{j+1} - 2f_j + f_{j-1}}{h^2} + O(h^2). \tag{2.9}$$

2.2 A General Technique for Construction of Finite Difference Schemes

A finite difference formula is characterized by the points at which the functional values are used and its order of accuracy. For example, the scheme in (2.9) uses the functional values at $j-1$, j, and $j+1$, and it is second-order accurate. Given a set of points to be used in a formula, called a stencil, it is desirable to construct the formula with the highest order accuracy that involves those points. There is a general procedure for constructing difference schemes that satisfies this objective; it is best described by an actual example. Suppose we want to construct the *most accurate* difference scheme that involves the functional values at points j, $j+1$, and $j+2$. In other words, given the restriction on the points involved, we ask for the highest order of accuracy that can be achieved. The desired finite difference formula can be written as

$$f'_j + \sum_{k=0}^{2} a_k f_{j+k} = O(?), \tag{2.10}$$

where a_k are the coefficients from the linear combination of Taylor series. These coefficients are to be determined so as to maximize the order of the scheme, which at this point is displayed by a question mark. We take the linear combination of the Taylor series for the terms in formula (2.10) using a convenient table shown below. The table displays the first four terms in the Taylor series expansion of the functional values in the first column.

TAYLOR TABLE

	f_j	f'_j	f''_j	f'''_j
f'_j	0	1	0	0
$a_0 f_j$	a_0	0	0	0
$a_1 f_{j+1}$	a_1	$a_1 h$	$a_1 \frac{h^2}{2}$	$a_1 \frac{h^3}{6}$
$a_2 f_{j+2}$	a_2	$2ha_2$	$a_2 \frac{(2h)^2}{2}$	$a_2 \frac{(2h)^3}{6}$

The left-hand side of (2.10) is the sum of the elements in the first column of the table; the first four terms of its right-hand side are the sum of the rows in the next four columns of the table, respectively. Thus, (2.10) can be constructed by summing the bottom four rows in the table:

$$f'_j + \sum_{k=0}^{2} a_k f_{j+k} = (a_0 + a_1 + a_2) f_j + (1 + a_1 h + 2ha_2) f'_j + \left(a_1 \frac{h^2}{2} + a_2 \frac{(2h)^2}{2} \right) f''_j + \left(a_1 \frac{h^3}{6} + a_2 \frac{(2h)^3}{6} \right) f'''_j + \cdots. \quad (2.11)$$

To get the highest accuracy, we must set as many of the low-order terms to zero as possible. We have three free coefficients; therefore, we can set the coefficients of the first three terms to zero:

$$\begin{aligned} a_0 + a_1 + a_2 &= 0 \\ a_1 h + 2ha_2 &= -1 \\ a_1 h^2/2 + 2a_2 h^2 &= 0. \end{aligned}$$

Solving these equations leads to

$$a_1 = -\frac{2}{h} \qquad a_2 = \frac{1}{2h} \qquad a_0 = \frac{3}{2h}.$$

Thus, the resulting (second-order) formula is obtained by substituting these values for the coefficients in (2.10), after a minor rearrangement we obtain

$$f'_j = \frac{-3f_j + 4f_{j+1} - f_{j+2}}{2h} + O(h^2). \quad (2.12)$$

The leading order truncation error is the first term on the right-hand side of (2.11) that we could not set to zero; substituting for a_1 and a_2, it becomes

$$\frac{h^2}{3} f'''_j.$$

Thus, the best we can do is a second-order formula, given the restriction that the formula is to involve the functional values at j, $j+1$, and $j+2$. It is interesting to note that the magnitude of the truncation error of this formula is twice that of the second-order central difference scheme (2.7).

EXAMPLE 2.1 Accuracy of Finite Difference Schemes

We will consider three different finite difference schemes and investigate their accuracy by varying the grid spacing, h. The first derivative of a known function f will be approximated and compared with the exact derivative. We take

$$f(x) = \frac{\sin x}{x^3}.$$

The specific schemes under consideration are the first-, second-, and fourth-order formulas given by (2.3), (2.7), and (2.8). These are numerically evaluated at $x = 4$ and the absolute values of the differences from the exact solution are plotted as a function of h in Figure 2.1. Since the approximation errors are proportional to powers of h, it is instructive to use a log–log plot to reveal the order of accuracy of the schemes. For each scheme, the curve representing the log |error| vs. log h is expected to be a straight line with its slope equal to the order of the scheme. The slopes of the curves in Figure 2.1 verify the order of each method.

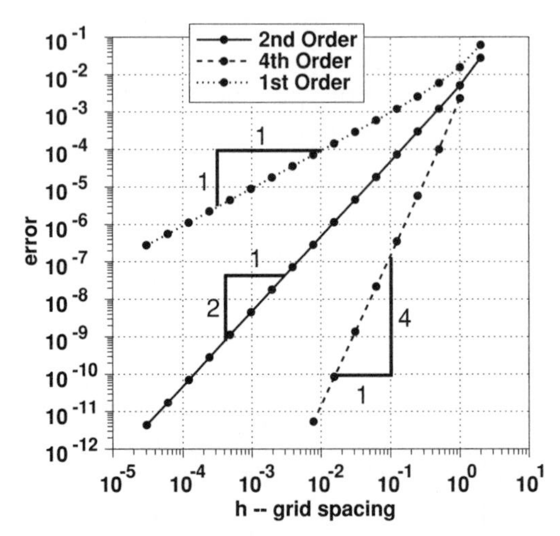

Figure 2.1 Truncation error vs. grid spacing for three finite difference schemes.

2.3 An Alternative Measure for the Accuracy of Finite Differences

Order of accuracy is the usual indicator of the accuracy of finite difference formulas; it tells us how mesh refinement improves the accuracy. For example, mesh refinement by a factor of 2 improves the accuracy of a second-order finite difference scheme by fourfold, and for a fourth-order scheme by a factor of 16.

Another method for measuring the order of accuracy that is sometimes more informative is the modified wavenumber approach. Here, we ask how well does a finite difference scheme differentiate a certain class of functions, namely sinusoidal functions. Sinusoidal functions are representative because Fourier series are often used to represent arbitrary functions. Of course, more points are required to adequately represent high-frequency sinusoidal functions and to differentiate them accurately. Given a set of points, or grid resolution, we are interested in knowing how well a finite difference scheme can differentiate the more challenging high-frequency sinusoidal functions. We expect that most differencing schemes would do well for the low-frequency, slowly varying functions. The solution of non-linear differential equations usually contains several frequencies and the modified wavenumber approach allows one to assess how well different components of the solution are represented.

To illustrate the procedure, consider a pure harmonic function of period L,

$$f(x) = e^{ikx},$$

where k is the wavenumber (or frequency) and can take on any of the following values

$$k = \frac{2\pi}{L}n, \qquad n = 0, 1, 2, \ldots, N/2.$$

With these values of k, each harmonic function would go through an integer number of periods in the domain. The exact derivative is

$$f' = ikf. \tag{2.13}$$

We now ask how accurately the second-order central finite difference scheme, for example, computes the derivative of f for different values of k. Let us discretize a portion of the x axis of length L with a uniform mesh,

$$x_j = \frac{L}{N}j, \qquad j = 0, 1, 2, \ldots, N-1.$$

On this grid, e^{ikx} ranges from a constant for $n = 0$, to a highly oscillatory function of period equal to two mesh widths for $n = N/2$. The finite difference approximation for the derivative is

$$\left.\frac{\delta f}{\delta x}\right|_j = \frac{f_{j+1} - f_{j-1}}{2h},$$

where $h = L/N$ is the mesh size and δ denotes the discrete differentiation operator. Substituting for $f_j = e^{ikx_j}$, we obtain

$$\left.\frac{\delta f}{\delta x}\right|_j = \frac{e^{i2\pi n(j+1)/N} - e^{i2\pi n(j-1)/N}}{2h} = \frac{e^{i2\pi n/N} - e^{-i2\pi n/N}}{2h} f_j.$$

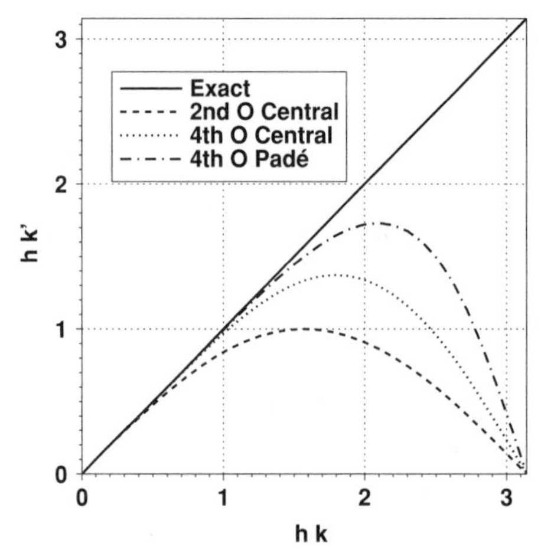

Figure 2.2 The modified wavenumbers for three finite difference schemes. h is the grid spacing. The Padé scheme is introduced in the next section.

Thus,

$$\left.\frac{\delta f}{\delta x}\right|_j = i\frac{\sin(2\pi n/N)}{h} f_j = ik' f_j$$

where

$$k' = \frac{\sin(2\pi n/N)}{h}. \tag{2.14}$$

The numerical approximation to the derivative is in the same form as the exact derivative in (2.13) except that k is replaced with k'. In analogy with (2.13), k' is called the modified wavenumber for the second-order central difference scheme. In an analogous manner, one can derive modified wavenumbers for any finite difference formula. A measure of accuracy of a finite difference scheme is provided by comparing the modified wavenumber k' with k. This comparison for three schemes is provided in Figure 2.2.

Note that the modified wavenumber in (2.14) (which is shown by the dash line in Figure 2.2) is in good agreement with the exact wavenumber at small values of k. This is expected because for small values of k, f is slowly varying and the finite difference scheme is sufficiently accurate for numerical differentiation. For higher values of k, however, f varies rapidly in the domain, and the finite difference scheme provides a poor approximation for its derivative. Although more accurate finite difference schemes provide better approximations at higher wavenumbers, the accuracy is always better for low wavenumbers compared to that for high wavenumbers. Similarly, we can assess the accuracy of any formula for a higher derivative using the modified wavenumber approach. For example, since the exact second derivative of the harmonic function is $-k^2 \exp(ikx)$, one can compare the modified wavenumber of a finite difference scheme for the second derivative, now labeled k'^2, with k^2. As for the first derivative, a typical k'^2h^2

vs. kh diagram shows better accuracy for small wavenumbers (see Exercise 6). It also turns out that the second-derivative finite difference formulas usually show better accuracy at the high wavenumbers than the first-derivative formulas.

2.4 Padé Approximations

The Taylor series procedure for obtaining the most accurate finite difference formula, given the functional values at certain points, can be generalized by inclusion of the derivatives at the neighboring grid points in the formula. For example, we can ask for the most accurate formula that includes f'_j, f'_{j+1}, and f'_{j-1} in addition to the functional values f_j, f_{j+1}, and f_{j-1}. That is, instead of (2.10), we would write

$$f'_j + a_0 f_j + a_1 f_{j+1} + a_2 f_{j-1} + a_3 f'_{j+1} + a_4 f'_{j-1} = O(?) \qquad (2.15)$$

and our task is then to find the five coefficients $a_0, a_1, \ldots, a_4$ to maximize the order of this approximation. Before worrying about how to use (2.15) for numerical differentiation, let us find the coefficients. We follow the Taylor table procedure for the functional values as well as derivatives appearing in (2.15). The Taylor table is

TAYLOR TABLE FOR A PADÉ SCHEME

	f_j	f'_j	f''_j	f'''_j	$f_j^{(iv)}$	$f_j^{(v)}$
f'_j	0	1	0	0	0	0
$a_0 f_j$	a_0	0	0	0	0	0
$a_1 f_{j+1}$	a_1	$a_1 h$	$a_1 \frac{h^2}{2}$	$a_1 \frac{h^3}{6}$	$a_1 \frac{h^4}{24}$	$a_1 \frac{h^5}{120}$
$a_2 f_{j-1}$	a_2	$-a_2 h$	$a_2 \frac{h^2}{2}$	$-a_2 \frac{h^3}{6}$	$a_2 \frac{h^4}{24}$	$-a_2 \frac{h^5}{120}$
$a_3 f'_{j+1}$	0	a_3	$a_3 h$	$a_3 \frac{h^2}{2}$	$a_3 \frac{h^3}{6}$	$a_3 \frac{h^4}{24}$
$a_4 f'_{j-1}$	0	a_4	$-a_4 h$	$a_4 \frac{h^2}{2}$	$-a_4 \frac{h^3}{6}$	$a_4 \frac{h^4}{24}$

As before, we now sum all the rows and set as many of the lower order terms to zero as possible. We have five coefficients and can set the sum of the entries in columns 2 to 6 to zero. The linear equations for the coefficients are

$$\begin{aligned}
a_0 + a_1 + a_2 &= 0 \\
a_1 h - a_2 h + a_3 + a_4 &= -1 \\
a_1 \frac{h^2}{2} + a_2 \frac{h^2}{2} + a_3 h - a_4 h &= 0 \\
a_1 \frac{h}{3} - a_2 \frac{h}{3} + a_3 + a_4 &= 0 \\
a_1 \frac{h}{4} + a_2 \frac{h}{4} + a_3 - a_4 &= 0.
\end{aligned}$$

The solution of this system is

$$a_0 = 0 \qquad a_1 = -\frac{3}{4h} \qquad a_2 = \frac{3}{4h} \qquad a_3 = a_4 = \frac{1}{4}.$$

Substitution into column 7 and (2.15) and some rearrangement leads to the following Padé formula for numerical differentiation:

$$f'_{j+1} + f'_{j-1} + 4f'_j = \frac{3}{h}(f_{j+1} - f_{j-1}) + \frac{h^4}{30}f^v_j, \tag{2.16}$$

where $j = 1, 2, 3, \ldots, n-1$.

This is a tridiagonal system of equations for f'_j. There are $n-1$ equations for $n+1$ unknowns. To get the additional equations, special treatment is required near the boundaries. Usually, lower order one-sided difference formulas are used to approximate f'_0 and f'_n. For example, the following third-order formulas provide the additional equations that would complete the set given by (2.16)

$$\begin{aligned} f'_0 + 2f'_1 &= \frac{1}{h}\left(-\frac{5}{2}f_0 + 2f_1 + \frac{1}{2}f_2\right) \\ f'_n + 2f'_{n-1} &= \frac{1}{h}\left(\frac{5}{2}f_n - 2f_{n-1} - \frac{1}{2}f_{n-2}\right). \end{aligned} \tag{2.17}$$

In matrix form, (2.16) and (2.17) are written as

$$\begin{bmatrix} 1 & 2 & 0 & 0 & 0 & \cdots & 0 \\ 1 & 4 & 1 & 0 & 0 & \cdots & 0 \\ 0 & 1 & 4 & 1 & 0 & \cdots & 0 \\ \vdots & \vdots & \ddots & \ddots & \ddots & \vdots & \vdots \\ \vdots & \vdots & \vdots & \ddots & \ddots & \ddots & \vdots \\ 0 & 0 & 0 & \ldots & 1 & 4 & 1 \\ 0 & 0 & 0 & 0 & \ldots & 2 & 1 \end{bmatrix} \begin{bmatrix} f'_0 \\ f'_1 \\ f'_2 \\ \vdots \\ \vdots \\ f'_{n-1} \\ f'_n \end{bmatrix} = \frac{1}{h} \begin{bmatrix} -\frac{5}{2}f_0 + 2f_1 + \frac{1}{2}f_2 \\ 3(f_2 - f_0) \\ 3(f_3 - f_1) \\ \vdots \\ \vdots \\ 3(f_n - f_{n-2}) \\ \frac{5}{2}f_n - 2f_{n-1} - \frac{1}{2}f_{n-2} \end{bmatrix}. \tag{2.18}$$

In choosing the boundary schemes, we consider two factors. First, in order to avoid writing a special code to solve the system of equations, the bandwidth of the matrix should not be increased. For example, the boundary scheme in (2.18) preserves the tridiagonal structure of the matrix which allows one to use a standard tridiagonal solver. Second, the boundary stencil should not be wider than the interior stencil. For example, if the interior stencil at x_1 involves only the functional and derivative values at x_0, x_1, and x_2, the boundary stencil should not include x_3. This constraint is derived from certain considerations in numerical solution of differential boundary value problems using finite differences (Chapter 4). The same constraint also applies to high-order standard non-Padé type schemes. For this reason, the order of the boundary scheme is usually lower than that of the interior scheme. However, there is substantial evidence from numerical tests that the additional errors due to a lower order boundary scheme are confined to the points near the boundaries.

EXAMPLE 2.2 Padé Differentiation Using a Lower Order Boundary Scheme

We will use the fourth-order Padé scheme (2.16) and the third-order boundary schemes given by (2.17) to differentiate

$$f(x) = \sin 5x \qquad 0 \le x \le 3.$$

Fifteen uniformly spaced points are used. The result is plotted in Figure 2.3. Although relatively few grid points are used, the Padé scheme is remarkably accurate. Note that the main discrepancies are near boundaries where lower order schemes are used.

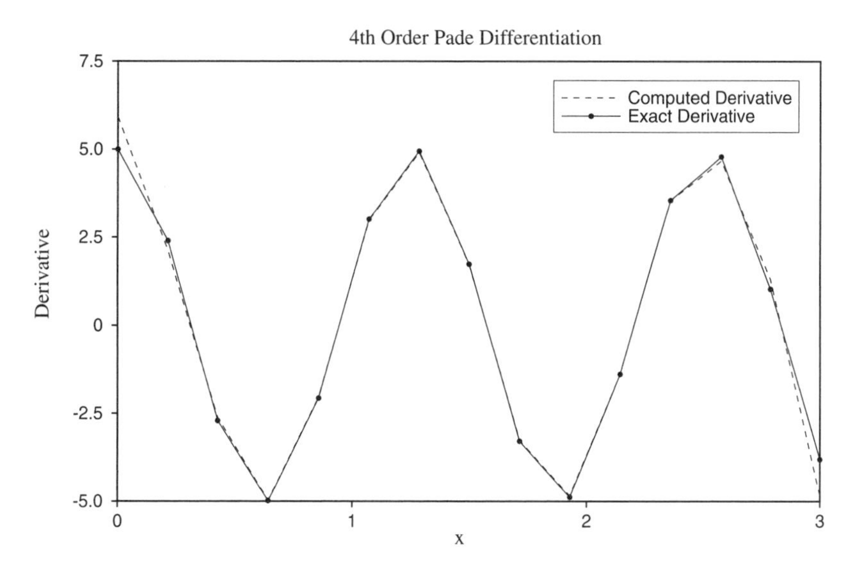

Figure 2.3 Computed derivative of the function in Example 2.2 using a fourth-order Padé scheme and exact derivative. The symbols mark the uniformly spaced grid points.

Note that despite its high order of accuracy, the Padé scheme (2.16) is compact; that is, it requires information only from the neighboring points, $j + 1$ and $j - 1$. Furthermore, as can be seen from Figure 2.1, this scheme has a more accurate modified wavenumber than the standard fourth-order scheme given by (2.8). Padé schemes are global in the sense that to obtain the derivative at a point, the functional values at all the points are required; one either gets the derivatives at all the points or none at all.

Padé schemes can also be easily constructed for higher derivatives. For example, for the three-point central stencil the following fourth-order formula can be derived using the Taylor table approach:

$$\frac{1}{12}f''_{i-1} + \frac{10}{12}f''_i + \frac{1}{12}f''_{i+1} = \frac{f_{i+1} - 2f_i + f_{i-1}}{h^2}. \tag{2.19}$$

2.5 Non-Uniform Grids

Often the function f varies rapidly in a part of the domain, and it has a mild variation elsewhere. In computationally intensive applications, it is considered wasteful to use a fine grid capable of resolving the rapid variations of f everywhere in the domain. One should use a non-uniform grid spacing. In some problems, such as boundary layers in fluid flow problems, the regions of rapid variations are known a priori, and grid points can be clustered where needed. There are also (adaptive) techniques that estimate the grid requirements as the solution progresses and place additional grid points in the regions of rapid variations. For now, we will just concern ourselves with finite differencing on non-uniformly spaced meshes.

Typical finite difference formulas for the first and second derivatives are

$$f'_j = \frac{f_{j+1} - f_{j-1}}{x_{j+1} - x_{j-1}} \tag{2.20}$$

and

$$f''_j = 2\left[\frac{f_{j-1}}{h_j(h_j + h_{j+1})} - \frac{f_j}{h_j h_{j+1}} + \frac{f_{j-1}}{h_{j+1}(h_j + h_{j+1})}\right], \tag{2.21}$$

where $h_j = x_j - x_{j-1}$. Finite difference formulas for non-uniform meshes generally have a lower order of accuracy than their counterparts with the same stencil but defined for uniform meshes. For example, (2.21) is strictly a first-order approximation whereas its counterpart on a uniform mesh (2.9) is second-order accurate. The lower accuracy is due to reduced cancellations in Taylor series expansions because of the lack of symmetry in the meshes.

An alternative to the cumbersome derivation of finite difference formulas on non-uniform meshes is to use a coordinate transformation. One may transform the independent variable to another coordinate system that is chosen to account for local variations of the solution. Uniform mesh spacing in the new coordinate system would correspond to non-uniform mesh spacing in the original (physical) coordinate (see Figure 2.4). For example, the transformation

$$\zeta = \cos^{-1} x$$

transforms $0 \leq x \leq 1$ to $0 \leq \zeta \leq \frac{\pi}{2}$. Uniform spacing in ζ, given by

$$\zeta_j = \frac{\pi}{2N} j \qquad j = 0, 1, 2, \ldots, N,$$

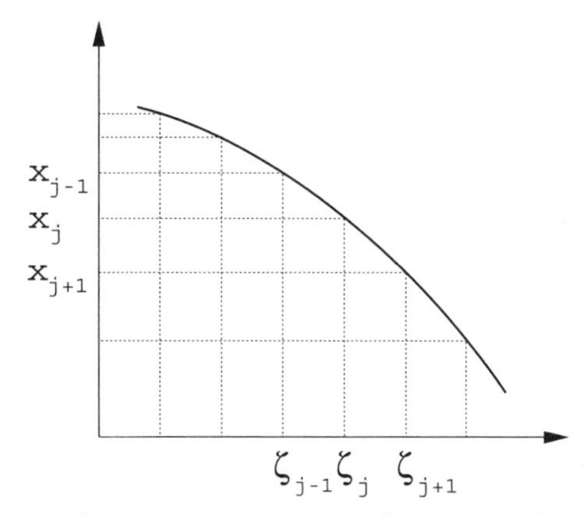

Figure 2.4 Uniform mesh spacing in ζ corresponds to non-uniform mesh spacing in x.

corresponds to a very fine mesh spacing near $x = 1$ and a coarse mesh near $x = 0$. In general, for the transformation

$$\zeta = g(x)$$

we use the chain rule to transform the derivatives to the new coordinate system

$$\frac{df}{dx} = \frac{d\zeta}{dx}\frac{df}{d\zeta} = g'\frac{df}{d\zeta} \tag{2.22}$$

$$\frac{d^2 f}{dx^2} = \frac{d}{dx}\left[g'\frac{df}{d\zeta}\right] = g''\frac{df}{d\zeta} + (g')^2\frac{d^2 f}{d\zeta^2}. \tag{2.23}$$

Finite difference approximations for uniform meshes are then used to approximate $df/d\zeta$ and $d^2 f/d\zeta^2$.

EXAMPLE 2.3 Calculation of Derivatives on a Non-uniform Mesh

Let f be a certain function defined on the grid points

$$x_j = \tanh^{-1}\zeta_j \qquad \text{where } \zeta_j = 0.9\left(\frac{2j}{N} - 1\right), \quad j = 0, \ldots, N.$$

The value of f at x_j is denoted by f_j. The x mesh is non-uniform and was constructed to have clustered points in the middle of the domain where f is supposed to exhibit rapid variations. The x mesh is shown versus the ζ mesh in Figure 2.5 for $N = 18$.

From (2.22), the first derivative of f at x_j is

$$\left.\frac{df}{dx}\right|_{x_j} = g'(x_j)\left.\frac{df}{d\zeta}\right|_{\zeta_j}.$$

The central difference approximation to

$$\left.\frac{df}{d\zeta}\right|_{\zeta_j}$$

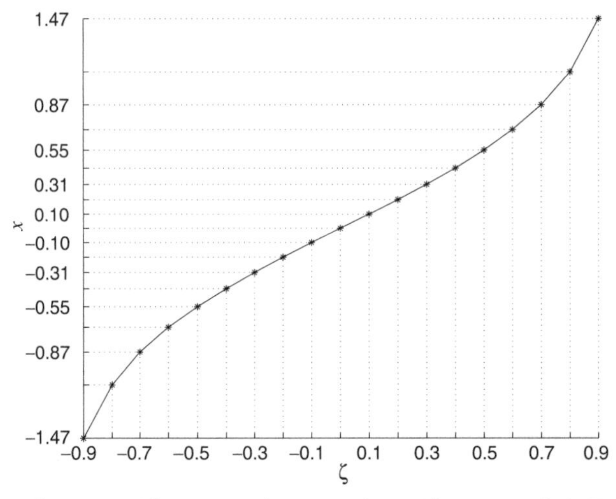

Figure 2.5 The non-uniform x mesh versus the uniform ζ mesh in Example 2.3.

is simply $(f_{j+1} - f_{j-1})/(2\Delta\zeta)$. In order to see this, let $y_1(x)$ describe f as a function of x. Then f as a function of ζ is given by $f = y_1(x) = y_1(g^{-1}(\zeta)) = y_2(\zeta)$, where y_2 is the composition of y_1 and g^{-1}. Thus

$$\left.\frac{df}{d\zeta}\right|_{\zeta_j} \approx \frac{y_2(\zeta_{j+1}) - y_2(\zeta_{j-1})}{2\Delta\zeta} = \frac{y_1(x_{j+1}) - y_1(x_{j-1})}{2\Delta\zeta} = \frac{f_{j+1} - f_{j-1}}{2\Delta\zeta}$$

and

$$\left.\frac{df}{dx}\right|_{x_j} \approx \coth(x_j)\frac{f_{j+1} - f_{j-1}}{2\Delta\zeta}.$$

An expression for the second derivative of f is obtained similarly.

These numerical derivatives are valid for $j = 1, \ldots, N-1$. Derivatives at $j = 0$ and N are obtained by using one-sided difference approximations to $df/d\zeta$ and $d^2f/d\zeta^2$.

EXERCISES

1. Consider the central finite difference operator $\delta/\delta x$ defined by

$$\frac{\delta u_n}{\delta x} = \frac{u_{n+1} - u_{n-1}}{2h}.$$

(a) In calculus we have

$$\frac{duv}{dx} = u\frac{dv}{dx} + v\frac{du}{dx}.$$

Does the following analogous finite difference expression hold?

$$\frac{\delta(u_n v_n)}{\delta x} = u_n\frac{\delta v_n}{\delta x} + v_n\frac{\delta u_n}{\delta x}.$$

(b) Show that

$$\frac{\delta(u_n v_n)}{\delta x} = \bar{u}_n \frac{\delta v_n}{\delta x} + \bar{v}_n \frac{\delta u_n}{\delta x}$$

where an overbar indicates average over the nearest neighbors,

$$\bar{u}_n = \frac{1}{2}(u_{n+1} + u_{n-1}).$$

(c) Show that

$$\phi \frac{\delta \psi}{\delta x} = \frac{\delta}{\delta x} \bar{\phi} \psi - \overline{\psi \frac{\delta \phi}{\delta x}}.$$

(d) Derive a finite difference formula for the second-derivative operator that is obtained from two applications of the first-derivative finite difference operator. Compare the leading error term of this formula and the popular second-derivative formula

$$\frac{u_{n+1} - 2u_n + u_{n-1}}{h^2}.$$

Use both schemes to calculate the second derivative of $\sin 5x$ at $x = 1.5$. Plot the absolute values of the errors as a function of h on a log–log plot similar to Figure 2.1. Use $10^{-4} \leq h \leq 10^0$. Discuss your plot.

2. Find the most accurate formula for the first derivative at x_i utilizing known values of f at x_{i-1}, x_i, x_{i+1}, and x_{i+2}. The points are uniformly spaced. Give the leading error term and state the order of the method.

3. Verify that the modified wavenumber for the fourth-order Padé scheme for the first derivative is

$$k' = \frac{3\sin(k\Delta)}{\Delta(2 + \cos(k\Delta))}.$$

4. A general Padé type boundary scheme (at $i = 0$) for the first derivative which does not alter the tridiagonal structure of the matrix in (2.16) can be written as

$$f_0' + \alpha f_1' = \frac{1}{h}(af_0 + bf_1 + cf_2 + df_3).$$

(a) Show that requiring this scheme to be at least third-order accurate would constrain the coefficients to

$$a = -\frac{11 + 2\alpha}{6}, \quad b = \frac{6 - \alpha}{2}, \quad c = \frac{2\alpha - 3}{2}, \quad d = \frac{2 - \alpha}{6}.$$

Which value of α would you choose and why?

(b) Find all the coefficients such that the scheme would be fourth-order accurate.

5. Modified wavenumbers for non-central finite difference schemes are complex. Derive the modified wavenumber for the down-wind scheme given by (2.12). Plot its real and imaginary parts separately and discuss your results.

6. *Modified wavenumber for second-derivative operators.*
Recall that the second derivative of $f = \exp(ikx)$ is $-k^2 f$. Application of a finite difference operator for second derivative to f would lead to $-k'^2 f$, where k'^2 is the 'modified wavenumber' for the second-derivative. The modified wavenumber method for assessing the accuracy of second-derivative finite difference formulas is then to compare the corresponding k'^2 with k^2 in a plot such as in Figure 2.1 (but now, $k'^2 h^2$ and $k^2 h^2$ vs. kh, $0 \leq kh \leq \pi$).
   (a) Use the modified wavenumber analysis to assess the accuracy of the central difference formula
$$f_j'' = \frac{f_{j+1} - 2f_j + f_{j-1}}{h^2}.$$
   (b) Use Taylor series to show that the Padé formula given by (2.19) is fourth-order accurate.
   (c) Use the modified wavenumber analysis to compare the schemes in (a) and (b). (Hint: To derive modified wavenumbers for Padé type schemes, replace f_j'' with $-k'^2 \exp(ikx_j)$, etc.)

FURTHER READING

Dahlquist, G., and Björck, Å. *Numerical Methods*. Prentice-Hall, 1974, Chapter 7.
Lapidus, L., and Pinder, George F. *Numerical Solution of Partial Differential Equations in Science and Engineering*. Wiley, 1982, Chapter 2.

3

Numerical Integration

Generally, numerical methods for integration or quadrature are needed more in practice than finite difference formulae for differentiation. The reason is that while differentiation is always possible to do analytically (even though it might sometimes be tedious) some integrals are difficult or impossible to do analytically. Therefore, we often refer to tables to evaluate non-trivial integrals. In this chapter we will introduce numerical methods that are used for evaluation of definite integrals that cannot be found in the tables; that is, they are impossible or too tedious to do analytically. Some of the elementary methods that are introduced can also be used to evaluate integrals where the integrand is only defined on a discrete grid or in tabular form.

Throughout the chapter, we will discuss methods for evaluation of the definite integral of the function f in the interval $[a, b]$,

$$I = \int_a^b f(x)\, dx.$$

We will assume that the functional values are known on a set of discrete points, $x_0 = a, x_1, x_2, \ldots, x_n = b$. If f is known analytically, the user or the algorithm would determine the location of the discrete points x_j. On the other hand if the data on f are available only in tabular form, then the locations of the grid points are fixed a priori and only a limited class of methods are applicable.

3.1 Trapezoidal and Simpson's Rules

For one interval, $x_i \le x \le x_{i+1}$, the trapezoidal rule is given by

$$\int_{x_i}^{x_{i+1}} f(x)\, dx \approx \frac{\Delta_x}{2}(f_i + f_{i+1}) \tag{3.1}$$

where $\Delta_x = x_{i+1} - x_i$. The geometrical foundation of this formula is that the function f in the interval is approximated by a straight line passing through the end points, and the area under the curve in the interval is approximated by

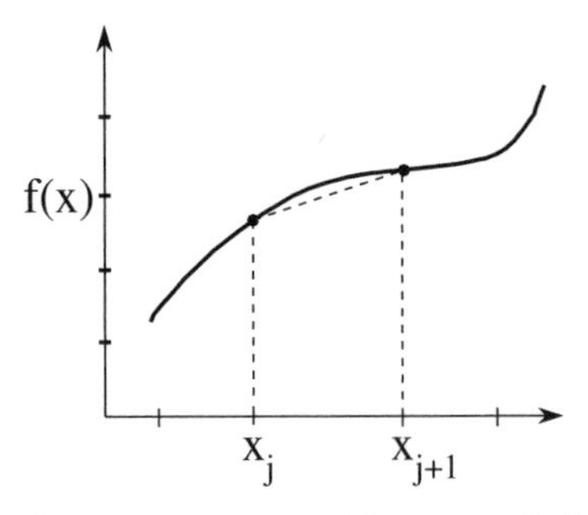

Figure 3.1 Trapezoidal rule; approximating f by a straight line between x_j and x_{j+1}.

the area of the resulting trapezoid (see Figure 3.1). For the entire interval $[a, b]$ the trapezoidal rule is obtained by adding the integrals over all sub-intervals:

$$I \approx \Delta_x \left(\frac{1}{2} f_0 + \frac{1}{2} f_n + \sum_{j=1}^{n-1} f_j \right), \tag{3.2}$$

where *uniform spacing* is assumed.

If we approximate f in each interval by a parabola rather than a straight line, then the resulting quadrature formula is known as *Simpson's rule*. To uniquely define a parabola as a fitting function, it must pass through three points (or two intervals). Thus, Simpson's formula for the integral from x_j to x_{j+2} is given by

$$\int_{x_j}^{x_{j+2}} f(x)\,dx \approx \frac{\Delta_x}{3} [f(x_j) + 4f(x_{j+1}) + f(x_{j+2})]. \tag{3.3}$$

Similarly, Simpson's rule for the entire domain is given by

$$I \approx \frac{\Delta_x}{3} \left(f_0 + f_n + 4 \sum_{\substack{j=1 \\ j=\text{odd}}}^{n-1} f_j + 2 \sum_{\substack{j=2 \\ j=\text{even}}}^{n-2} f_j \right). \tag{3.4}$$

Note that in order to use Simpson's rule for the entire interval of integration, the total number of points $(n + 1)$ must be odd (even number of panels).

Before we discuss the accuracy of these formulae, notice that they both can be written in the compact form:

$$I = \int_a^b f(x)\,dx \approx \sum_{i=0}^{n} w_i f(x_i) \tag{3.5}$$

where w_i are the weights. For example, for the trapezoidal rule $w_0 = w_n = \frac{h}{2}$ and $w_i = h$ for $i = 1, 2, \ldots, n - 1$.

3.2 Error Analysis

We will now establish the accuracy of these formulas using Taylor series expansions. It turns out that it is easier to build our analysis around the so-called

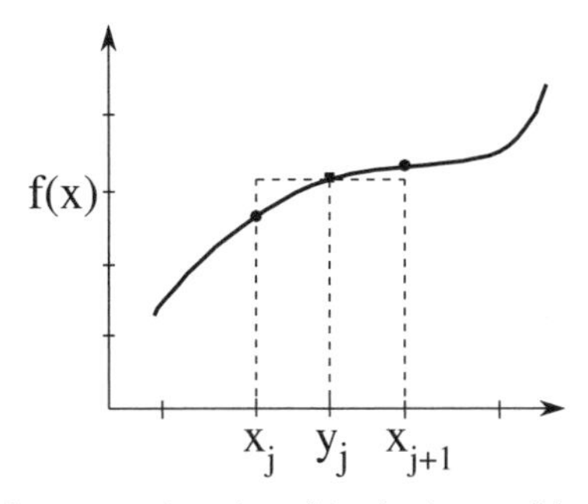

Figure 3.2 Rectangle rule; approximating f in the interval between x_j and x_{j+1} by its value at the midpoint.

rectangle (or midpoint) rule of integration; the order of accuracy of the trapezoidal and Simpson rules are then easily derived from that of the rectangle rule.

Consider the rectangle rule (Figure 3.2) for the interval $[x_i, x_{i+1}]$,

$$\int_{x_i}^{x_{i+1}} f(x)\,dx \approx h_i f(y_i), \tag{3.6}$$

where $y_i = (x_i + x_{i+1})/2$ is the midpoint of the interval $[x_i, x_{i+1}]$ and h_i is its width. Let's replace the integrand with its Taylor series about y_i

$$f(x) = f(y_i) + (x - y_i)f'(y_i) + \frac{1}{2}(x - y_i)^2 f''(y_i) + \frac{1}{6}(x - y_i)^3 f'''(y_i) + \cdots.$$

Substitution in (3.6) leads to

$$\int_{x_i}^{x_{i+1}} f(x)\,dx = h_i f(y_i) + \frac{1}{2}(x - y_i)^2 \Big|_{x_i}^{x_{i+1}} f'(y_i) + \frac{1}{6}(x - y_i)^3 \Big|_{x_i}^{x_{i+1}} f''(y_i) + \cdots.$$

All the terms with even powers of $(x - y_i)$ vanish, and we obtain

$$\int_{x_i}^{x_{i+1}} f(x)\,dx = h_i f(y_i) + \frac{h_i^3}{24} f''(y_i) + \frac{1}{1920} h_i^5 f^{(iv)}(y_i) + \cdots. \tag{3.7}$$

Thus, for one interval, the rectangle rule is third-order accurate.

Now let us perform an error analysis for the trapezoidal rule. Consider the Taylor series expansions for the functional values appearing on the right-hand side of (3.1):

$$f(x_i) = f(y_i) - \frac{1}{2}h_i f'(y_i) + \frac{1}{8}h_i^2 f_i''(y_i) - \frac{1}{48}h_i^3 f'''(y_i) + \cdots$$

$$f(x_{i+1}) = f(y_i) + \frac{1}{2}h_i f'(y_i) + \frac{1}{8}h_i^2 f_i''(y_i) + \frac{1}{48}h_i^3 f'''(y_i) + \cdots.$$

Adding these two expressions and dividing by 2 yields,

$$\frac{f(x_i)+f(x_{i+1})}{2} = f(y_i) + \frac{1}{8}h_i^2 f''(y_i) + \frac{1}{384}h_i^4 f^{(iv)}(y_i) + \cdots.$$

Now we can use this expression to solve for $f(y_i)$ and then substitute it into (3.7)

$$\int_{x_i}^{x_{i+1}} f(x)\,dx = h_i \frac{f(x_i)+f(x_{i+1})}{2} - \frac{1}{12}h_i^3 f''(y_i) - \frac{1}{480}h_i^5 f^{(iv)}(y_i) + \cdots. \quad (3.8)$$

Thus, for one interval the trapezoidal rule is also third-order accurate, and its leading truncation error is twice in magnitude but has the opposite sign of the truncation error of the rectangle rule. This is a bit surprising since we would expect approximating a function in an interval by a straight line (which is the basis of the trapezoidal method) to be more accurate than approximating it by a horizontal line passing through the function at the midpoint of the interval. Apparently, error cancellations in evaluating the integral lead to higher accuracy for the rectangle rule.

To obtain the order of accuracy for approximating the integral for the *entire domain*, we can sum both sides of (3.8); assuming uniform spacing, i.e., $h_i = \Delta$, we will have

$$\begin{aligned} I = \int_a^b f(x)\,dx &\approx \sum_{i=0}^{n-1} \int_{x_i}^{x_{i+1}} f(x)\,dx \\ &= \frac{\Delta}{2}\left(f(a) + f(b) + 2\sum_{j=1}^{n-1} f_j\right) - \frac{\Delta^3}{12}\sum_{i=0}^{n-1} f''(y_i) \\ &\quad - \frac{\Delta^5}{480}\sum_{i=0}^{n-1} f^{(iv)}(y_i) + \cdots. \end{aligned} \quad (3.9)$$

Now, we will apply the mean value theorem of integral calculus to the summations. The mean value theorem states that for sufficiently smooth f there exists a point $\bar{x}$ in the interval $[a, b]$ such that

$$\sum_{i=0}^{n-1} f''(y_i) = n f''(\bar{x}).$$

Similarly, there is a point ξ in $[a, b]$, such that

$$\sum_{i=0}^{n-1} f^{(iv)}(y_i) = n f^{(iv)}(\xi).$$

Noting that $n = (b-a)/\Delta$ and using the results of the mean value theorem in

(3.9), we obtain

$$I = \int_a^b f(x)\,dx = \frac{\Delta}{2}\left[f(a) + f(b) + 2\sum_{j=1}^{n-1} f_j\right]$$
$$- (b-a)\frac{\Delta^2}{12}f''(\bar{x}) - (b-a)\frac{\Delta^4}{480}f^{(iv)}(\xi) + \cdots. \qquad (3.10)$$

Thus, the trapezoidal rule for the *entire interval* is *second-order* accurate. One can easily show that the Simpson's formula for one panel $[x_i, x_{i+2}]$ can be written as

$$S(f) = \frac{2}{3}R(f) + \frac{1}{3}T(f),$$

where $R(f)$ and $T(f)$ denote rectangle and trapezoidal rules, respectively, applied to the function f. Note that the midpoint of the interval $[x_i, x_{i+2}]$ is x_{i+1}. Using (3.7) and (3.8) (modified for the interval $[x_i, x_{i+2}]$) and the mean value theorem, we see that Simpson's rule is *fourth-order* accurate for the entire interval $[a, b]$.

3.3 Trapezoidal Rule with End-Correction

This rule is easily derived by simply substituting in (3.8) for $f''(y_i)$, the second-order central difference formula, $f''(y_i) = (f'_{i+1} - f'_i)/h_i + O(h_i^2)$:

$$I_i = h_i\frac{f_i + f_{i+1}}{2} - \frac{1}{12}h_i^3\frac{f'_{i+1} - f'_i}{h_i} + O(h_i^5).$$

Once again, to get a simple global integration formula, we will assume constant step size, $h_i = h =$ const, and sum over the entire interval

$$I = \frac{h}{2}\sum_{i=0}^{n-1}(f_i + f_{i+1}) - \frac{h^2}{12}\sum_{i=0}^{n-1}(f'_{i+1} - f'_i) + O(h^4).$$

Cancellations in the second summation on the right-hand side lead to

$$I = \frac{h}{2}\sum_{i=0}^{n-1}(f_i + f_{i+1}) - \frac{h^2}{12}(f'(b) - f'(a)) + O(h^4). \qquad (3.11)$$

Thus, the trapezoidal rule with end-correction is *fourth-order* accurate and can be readily applied without much additional work, provided that the derivatives of the integrand at the end points are known.

EXAMPLE 3.1 Quadrature

Consider the integral

$$\int_1^{\pi} \frac{\sin x}{2x^3}\,dx.$$

We will numerically evaluate this integral using the trapezoidal rule (3.3), Simpson's rule (3.4), and trapezoidal rule with end-correction (3.11). This integral has an analytical solution in terms of Si(x), sine integrals (see *Handbook of Mathematical Functions*, by Abramowitz & Stegun, p. 231), and may be numerically evaluated to an arbitrary degree of accuracy for use as an 'exact' solution, allowing us to evaluate our quadrature techniques. The results of the numerical calculations as well as percent errors† for the quadrature techniques are presented below for $N = 8$ and $N = 32$ panels in the integration. The 'exact' solution is $I = 0.1985572988\ldots$.

$N = 8$	**Result**	**% Error**
Trapezoidal	0.204304	2.894303
Simpson	0.198834	0.139596
End-Correct.	0.198476	0.040948

$N = 32$	**Result**	**% Error**
Trapezoidal	0.198921	0.183286
Simpson	0.198559	0.000661
End-Correct.	0.198557	0.000167

We see that the higher order Simpson's rule and trapezoidal with end-correction outperform the plain trapezoidal rule.

3.4 Romberg Integration and Richardson Extrapolation

Richardson extrapolation is a powerful technique for obtaining an accurate numerical solution of a quantity (e.g., integral, derivative, etc.) by combining two or more less accurate solutions. The essential ingredient for application of the technique is knowledge of the form of the truncation error of the basic numerical method used. We shall demonstrate application of the Richardson extrapolation by using it to improve the accuracy of the integral

$$I = \int_a^b f(x)\, dx$$

with the trapezoidal rule as the basic numerical method. This algorithm is known as the *Romberg integration*.

† The percent error (% error) is the absolute value of the truncation error divided by the exact solution and multiplied by 100:

$$\%\ \text{error} = \left| \frac{\text{exact solution} - \text{numerical solution}}{\text{exact solution}} \right| \times 100.$$

From our error analysis for the trapezoidal rule (3.10), we have

$$I = \frac{h}{2}\left[f(a) + f(b) + 2\sum_{j=1}^{n-1} f_j\right] + c_1 h^2 + c_2 h^4 + c_3 h^6 + \cdots. \tag{3.12}$$

Let the trapezoidal approximation with uniform mesh of size h be denoted by $\tilde{I}_1$

$$\tilde{I}_1 = \frac{h}{2}\left[f(a) + f(b) + 2\sum_{j=1}^{n-1} f_j\right]. \tag{3.13}$$

The exact integral and the trapezoidal expression are related by

$$\tilde{I}_1 = I - c_1 h^2 - c_2 h^4 - c_3 h^6 - \cdots. \tag{3.14}$$

Now, suppose we evaluate the integral with half the step size $h_1 = h/2$. Let's call this estimate $\tilde{I}_2$

$$\tilde{I}_2 = I - c_1 \frac{h^2}{4} - c_2 \frac{h^4}{16} - c_3 \frac{h^6}{64} - \cdots. \tag{3.15}$$

We can eliminate $O(h^2)$ terms by taking a linear combination of (3.14) and (3.15) to obtain

$$\tilde{I}_{12} = \frac{4\tilde{I}_2 - \tilde{I}_1}{3} = I + \frac{1}{4} c_2 h^4 + \frac{5}{16} c_3 h^6 + \cdots. \tag{3.16}$$

This is a *fourth-order* approximation for I. In fact, (3.16) is a rediscovery of Simpson's rule. We have combined two estimates of I to obtain a more accurate estimate; this procedure is called the Richardson extrapolation and can be repeated to obtain still higher accuracy.

Let's evaluate I with $h_2 = h_1/2 = h/4$; we obtain

$$\tilde{I}_3 = I - c_1 \frac{h^2}{16} - c_2 \frac{h^4}{256} - c_3 \frac{h^6}{4096} - \cdots. \tag{3.17}$$

To get another fourth-order estimate, we will combine $\tilde{I}_3$ with $\tilde{I}_2$:

$$\tilde{I}_{23} = \frac{4\tilde{I}_3 - \tilde{I}_2}{3} = I + \frac{1}{64} c_2 h^4 + \frac{5}{1024} c_3 h^6 + \cdots. \tag{3.18}$$

Now that we have two fourth-order estimates, we can combine them and eliminate the $O(h^4)$ terms. Elimination of the $O(h^4)$ terms between (3.16) and (3.18) results in a *sixth-order accurate* formula. This process can be continued indefinitely. The essence of the Romberg integration algorithm just described is illustrated in the following diagram. In typical Romberg integration subroutines, the user specifies an error tolerance, and the algorithm uses the Richardson extrapolation as many times as necessary to achieve it.

$$\begin{array}{ccc} O(h^2) & O(h^4) & O(h^6) \\ \tilde{I}_1 & & \\ \tilde{I}_2 & \text{Eqn. (3.16)} & \\ \tilde{I}_3 & \text{Eqn. (3.18)} & \rightarrow \end{array}$$

EXAMPLE 3.2 Romberg Integration

We will numerically evaluate the integral from Example 3.1 using the Romberg integration. The basis for our integration will be the trapezoidal rule. The integration will be set to automatically stop when the solution varies less than 0.1% between levels – we may thus specify how accurate we wish our solution to be. The table below shows the Romberg integration in progress. The first column indicates the number of panels used to compute the integral using the trapezoidal rule.

2	$\tilde{I}_1 = 0.278173$			
4	$\tilde{I}_2 = 0.220713$	0.201560		
8	$\tilde{I}_3 = 0.204304$	0.198834	0.198653	
16	$\tilde{I}_4 = 0.200009$	0.198578	0.198560	0.198559

The % error of this calculation is 0.00074. We see that using only a second-order method as a basis we are able to generate an $O(h^8)$ method and a 0.00074% error at the cost of only 17 function evaluations.

3.5 Adaptive Quadrature

Often it is wasteful to use the same mesh size everywhere in the interval of integration $[a, b]$. The major cost of numerical integration is the number of function evaluations required, which is obviously related to the number of mesh points used. Thus, to reduce the computational effort, one should use a fine mesh only in regions of rapid functional variation and a coarser mesh where the integrand is varying slowly. Adaptive quadrature techniques automatically determine panel sizes in various regions so that the computed result meets some prescribed accuracy requirement supplied by the user. That is, with the minimum number of function evaluations, we would like a numerical estimate $\tilde{I}$ of the integral such that

$$\left| \tilde{I} - \int_a^b f(x)\,dx \right| \leq \epsilon$$

where ϵ is the error tolerance provided by the user.

To demonstrate the technique, we will use Simpson's rule as the base method. Let's divide the interval $[a, b]$ into subintervals $[x_i, x_{i+1}]$. Divide this interval into two panels and use Simpson's rule to obtain

$$S_i = \frac{h_i}{6}\left[f(x_i) + 4f\left(x_i + \frac{h_i}{2}\right) + f(x_i + h_i)\right].$$

Now, divide the interval into four panels, and obtain another estimate for the integral

$$S_i^{(2)} = \frac{h_i}{12}\left[f(x_i) + 4f\left(x_i + \frac{h_i}{4}\right) + 2f\left(x_i + \frac{h_i}{2}\right) + 4f\left(x_i + \frac{3h_i}{4}\right) + f(x_i + h_i)\right].$$

The basic idea, as will be shown, is to compare the two approximations, S_i and $S_i^{(2)}$, and obtain an estimate for the accuracy of $S_i^{(2)}$. If the accuracy is acceptable, we will use $S_i^{(2)}$ for the interval and start working on the next interval; otherwise, the method further subdivides the interval. Let I_i denote the exact integral in $[x_i, x_{i+1}]$. From our error analysis we know that Simpson's rule is *locally* fifth-order accurate,

$$I_i - S_i = ch_i^5 f^{(iv)}\left(x_i + \frac{h_i}{2}\right) + \cdots \tag{3.19}$$

and for the refined interval, we simply add the two truncation errors

$$I_i - S_i^{(2)} = c\left(\frac{h_i}{2}\right)^5\left[f^{(iv)}\left(x_i + \frac{h_i}{4}\right) + f^{(iv)}\left(x_i + \frac{3h_i}{4}\right)\right] + \cdots.$$

Each of the terms in the bracket can be expanded in Taylor series about the point $(x_i + h_i/2)$:

$$f^{(iv)}\left(x_i + \frac{h_i}{4}\right) = f^{(iv)}\left(x_i + \frac{h_i}{2}\right) - \frac{h_i}{4}f^{(v)}\left(x_i + \frac{h_i}{2}\right) + \cdots$$

$$f^{(iv)}\left(x_i + \frac{3h_i}{4}\right) = f^{(iv)}\left(x_i + \frac{h_i}{2}\right) + \frac{h_i}{4}f^{(v)}\left(x_i + \frac{h_i}{2}\right) + \cdots.$$

Thus,

$$I_i - S_i^{(2)} = 2c\left(\frac{h_i}{2}\right)^5\left[f^{(iv)}\left(x_i + \frac{h_i}{2}\right)\right] + \cdots. \tag{3.20}$$

Subtracting (3.19) from (3.20), I_i drops out and we obtain

$$S_i^{(2)} - S_i = \frac{15}{16}ch_i^5 f^{(iv)}\left(x_i + \frac{h_i}{2}\right) + \cdots.$$

This is the key result, it states that the error in $S_i^{(2)}$, as given by (3.20), is about

$1/15$ of the difference between S_i and $S_i^{(2)}$. The good news is that this difference can be computed; it is simply the difference between two numerical estimates of the integral that we have already computed.

If the user-specified error tolerance for the entire interval is ϵ, the weighted tolerance for the interval $[x_i, x_{i+1}]$ is

$$\frac{h_i}{b-a}\epsilon.$$

Thus, the adaptive algorithm proceeds as follows: If

$$\frac{1}{15}\left|S_i^{(2)} - S_i\right| \le \frac{h_i}{b-a}\epsilon, \tag{3.21}$$

then $S_i^{(2)}$ is sufficiently accurate for the interval $[x_i, x_{i+1}]$, and we move on to the next interval. If condition (3.21) is not satisfied, the interval will be subdivided further.

This is the essence of adaptive quadrature programs. Similar methodology can be devised when other base methods such as the trapezoidal rule are used. As with the Richardson extrapolation, the knowledge of the truncation error can be used to obtain estimates for the accuracy of the numerical solution *without* knowing the exact solution.

EXAMPLE 3.3 Adaptive Quadrature

Consider the function

$$f(x) = 10e^{-50|x|} - \frac{0.01}{(x-0.5)^2 + 0.001} + 5\sin(5x).$$

The integral

$$I = \int_{-1}^{1} f(x)\,dx$$

has the exact value of -0.56681975015. When evaluated using the adaptive quadrature routine QUANC8†, with various error tolerances ϵ, the following values are obtained.

ϵ	**Integral**
10^{-2}	-0.45280954
10^{-3}	-0.53238036
10^{-4}	-0.56779547
10^{-5}	-0.56681371
10^{-6}	-0.56681977
10^{-7}	-0.56681974

† G. E. Forsythe, M. A. Malcolm, and C. B. Moler (1977), *Computer Methods for Mathematical Computations*. Englewood Cliffs, N.J.: Prentice Hall. QUANC8 is available on the World Wide Web; check, for example, `http://www.netlib.org/`.

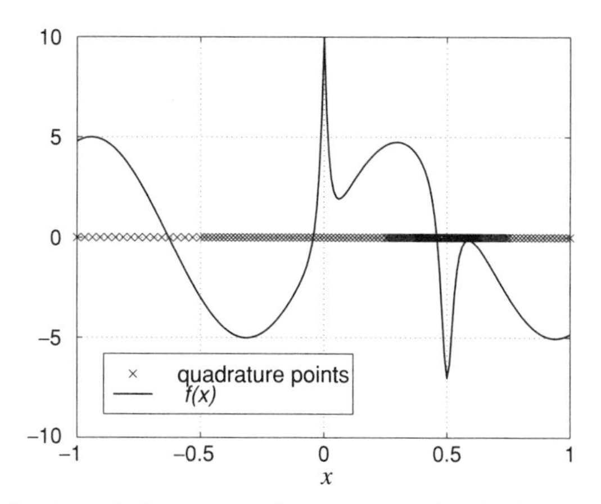

Figure 3.3 Distribution of adaptive quadrature points for the function in Example 3.3.

The quadrature points for the case $\epsilon = 10^{-5}$ are shown along with the function $f(x)$ in Figure 3.3. Note how the subroutine puts more points in regions where greater resolution was needed for evaluation of the integral.

3.6 Gauss Quadrature

Recall that any quadrature formula can be written as

$$I = \int_a^b f(x)\,dx = \sum_{i=0}^{n} w_i f(x_i). \tag{3.22}$$

If the function f is given analytically, we have two important choices to make. We have to select the location of the points x_i and the weights w_i. The main concept underpinning Gauss quadrature is to make these choices for optimal accuracy; the criterion for accuracy being the highest degree polynomial that can be integrated exactly. You can easily verify that the trapezoidal rule integrates a straight line exactly and Simpson's rule integrates a cubic exactly (see Exercise 4). As we will show below, Gauss quadrature integrates a polynomial of degree $2n+1$ exactly using only $n+1$ points, which is a remarkable achievement!

Let f be a polynomial of degree $2n+1$. Suppose we represent f by an nth-order Lagrange polynomial, P. Let $x_0, x_1, x_2, \ldots, x_n$ be the points on the x-axis where the function f is evaluated. Using Lagrange interpolation, we have:

$$P(x) = \sum_{j=0}^{n} f(x_j) L_j^{(n)}(x). \tag{3.23}$$

This representation is exact if f is a polynomial of degree n. Let F be a polynomial of degree $n+1$ with $x_0, x_1, \ldots, x_n$ as its roots,

$$F(x) = (x - x_0)(x - x_1)(x - x_2)\cdots(x - x_n).$$

The difference $f(x) - P(x)$ is a polynomial of degree $2n + 1$ that vanishes at $x_0, x_1, \ldots, x_n$ because P was constructed to pass through $f(x_0), f(x_1), \ldots, f(x_n)$ at the points $x_0, x_1, \ldots, x_n$. Thus, we can write the difference $f(x) - P(x)$ in the following form:

$$f(x) - P(x) = F(x) \sum_{l=0}^{n} q_l x^l.$$

Integrating this equation results in

$$\int f(x)\,dx = \int P(x)\,dx + \int F(x) \sum_{l=0}^{n} q_l x^l dx.$$

Suppose we demand that

$$\int F(x)x^{\alpha} dx = 0 \qquad \alpha = 0, 1, 2, 3, \ldots, n. \tag{3.24}$$

In principle we can choose $x_0, x_1, x_2, \ldots, x_n$ such that these $n + 1$ conditions are satisfied. Choosing the abscissa in this manner leads to the following expression for the integral:

$$\int f(x)\,dx = \int P(x)\,dx = \sum_{j=0}^{n} f(x_j) w_j,$$

where

$$w_j = \int L_j^{(n)}(x)\,dx \tag{3.25}$$

are the weights.

According to (3.24), F is a polynomial of degree $n + 1$ that is orthogonal to all polynomials of degree less than or equal to n. Points $x_0, x_1, \ldots, x_n$ are the zeros of this polynomial. These polynomials are called Legendre polynomials when x varies between -1 and 1. They are orthonormal, that is

$$\int_{-1}^{1} F_n(x) F_m(x)\,dx = \delta_{nm}$$

where

$$\delta_{nm} = \begin{cases} 0 & \text{if} \quad m \neq n \\ 1 & \text{if} \quad m = n, \end{cases}$$

and F_n is the Legendre polynomial of degree n. Their zeros are documented in mathematical tables (see *Handbook of Mathematical Functions*, by Abramowitz & Stegun) or in canned programs (see for example, *Numerical Recipes* by Press et al.). Having the zeros, the weights w_j can be readily computed, and they are also documented in the Gauss quadrature tables or obtained from canned programs. Many numerical analysis software libraries contain Gauss quadrature integration subroutines.

Note that we can always transform the interval $a \leq x \leq b$ into $-1 \leq \xi \leq 1$ by the transformation

$$x = \frac{b+a}{2} + \frac{b-a}{2}\xi.$$

Typically, to use Gauss–Legendre quadrature tables to evaluate the integral

$$\int_a^b f(x)\,dx,$$

one first changes the independent variable to ξ and obtains the weights w_i and the points on the abscissa, $\xi_0, \xi_2, \ldots, \xi_n$ from the tables (for the chosen n). The integral is then approximated by

$$\frac{b-a}{2}\sum_{j=0}^{n} f\left(\frac{b+a}{2} + \frac{b-a}{2}\xi_j\right) w_j. \tag{3.26}$$

Note that in the tables in Abramowitz & Stegun, n denotes the number of points, not $n+1$.

EXAMPLE 3.4 Integration Using Gauss–Legendre Quadrature

Consider the integral

$$\int_1^8 \frac{\log x}{x}\,dx.$$

The exact value is $\frac{1}{2}(\log 8)^2 = 2.1620386$. Suppose we evaluate this integral with five points using the Gauss–Legendre quadrature. The subroutine `gauleg` in *Numerical Recipes* gives the following points and weights in the interval, $1 \leq x \leq 8$:

i	x_i	w_i
1	1.3283706	0.8292441
2	2.6153574	1.6752003
3	4.5000000	1.9911112
4	6.3846426	1.6752003
5	7.6716294	0.8292441

Substituting these values into (3.22) results in the numerical estimate for the integral, $I \approx 2.165234$. The corresponding error is $\epsilon = 0.0032$ (0.15%) which is much better than the performance of the Simpson's rule with nine points (eight panels), i.e., $\epsilon = 0.013$ (0.6%). Gauss quadrature with nine points would result in $\epsilon = 0.000011$ (0.05%).

There are several Gauss quadrature procedures corresponding to other orthogonal polynomials. These polynomials are distinguished by the weight functions, W, used in their statement of orthogonality:

$$\int_a^b P_m(x)P_n(x)W(x)\,dx = \delta_{mn} \tag{3.27}$$

and the range $[a, b]$ over which the functions are orthogonal. For example, Hermite polynomials are orthogonal according to

$$\int_{-\infty}^{+\infty} e^{-x^2} H_m(x)H_n(x)\,dx = \delta_{mn}.$$

The Gauss–Hermite quadrature can be used to evaluate integrals of the form

$$I = \int_{-\infty}^{+\infty} e^{-x^2} f(x)\,dx \approx \sum_{i=0}^{n} w_i f(x_i). \tag{3.28}$$

This should lead to accurate results provided that f grows slower than e^{x^2} as $|x|$ approaches infinity.

EXAMPLE 3.5 Gauss Quadrature Based on Hermite Polynomials

Consider the integral

$$I = \int_{-\infty}^{+\infty} e^{-x^2} \cos x\,dx.$$

The exact value is 1.38038845. Suppose we use the Gauss–Hermite quadrature to evaluate the integral using seven nodes. A call to the `gauher` FORTRAN subroutine in *Numerical Recipes* produces the following abscissa and weights:

i	x_i	w_i
1	2.6519613	0.0009718
2	1.6735517	0.0545156
3	0.8162879	0.4256073
4	0.0000000	0.8102646
5	−0.8162879	0.4256073
6	−1.6735517	0.0545156
7	−2.6519613	0.0009718

Note that the weights rapidly vanish at higher values of $|x|$, this is probably why no more points are needed beyond $|x| = 2.652$. Substituting these values into (3.28) results in $I \approx 1.38038850$, which is in excellent agreement with the exact value.

Although Gauss quadrature is very powerful, it may not be cost effective for solution improvement. One improves the accuracy by adding additional points, which would involve additional function evaluations. Function evaluations are the major portion of the computational cost in numerical integration. In the case of Gauss quadrature, the new grid points generally do not include the old ones and therefore one needs to perform a complete new set of function evaluations.

In contrast, adaptive techniques and the Romberg integration do not discard the previous function evaluations but use them to improve the solution accuracy when additional points are added.

EXERCISES

1. Show that

$$\sum_{i=1}^{N-1} u_i \frac{\delta v}{\delta x}\bigg|_i = -\sum_{i=1}^{N-1} v_i \frac{\delta u}{\delta x}\bigg|_i + \text{boundary terms}.$$

   What are the boundary terms? Compare this discrete expression to the rule of integration by parts.

2. Using the error analysis for the trapezoidal and rectangle rules, show that Simpson's rule for integration over the entire interval is fourth-order accurate.
3. Explain why in Example 3.1, the trapezoidal rule with end-correction is slightly more accurate than the Simpson's rule.
4. Explain why the rectangle and trapezoidal rules can integrate a straight line exactly and the Simpson's rule can integrate a cubic exactly.
5. A common problem of mathematical physics is that of solving the Fredholm integral equation

$$f(x) = \phi(x) + \int_a^b K(x,t)\phi(t)\,dt,$$

   where the functions $f(x)$ and $K(x,t)$ are given and the problem is to obtain $\phi(x)$.

   (a) Describe a numerical method for solving this equation.

   (b) Solve the following equation

$$\phi(x) = \pi x^2 + \int_0^{\pi} 3(0.5\sin 3x - tx^2)\phi(t)\,dt.$$

   Compare to the exact solution $\phi(x) = \sin 3x$.

6. Describe a method for solving the Volterra integral equation

$$f(x) = \phi(x) + \int_a^x K(x,t)\phi(t)\,dt.$$

   Note that the upper limit of the integral is x. What is $\phi(a)$?

7. Consider the integral

$$\int_0^1 \left[\frac{100}{\sqrt{x+.01}} + \frac{1}{(x-0.3)^2+.001} - \pi\right] dx.$$

   (a) Numerically evaluate this integral using the trapezoidal rule with n panels of uniform length h. Make a log–log plot of the error (%) vs. n and discuss the accuracy of the method. Take $n = 8, 16, 32, \ldots$.

   (b) Repeat part (a) using the Simpson's rule and the trapezoidal rule with end-correction.

(c) Evaluate the integral using an adaptive method with various error tolerances (you may want to use the *Numerical Recipes* subroutine `odeint` or MATLAB's function `quad8`). How are the x points for function evaluations distributed? Plot the integrand showing the positions of its evaluations on the x axis.

8. Simpson's rule was used to find the value of the integral $I = \int_0^1 f(x)\,dx$. The results for two different step sizes are given in the table below

h	I
0.2	12.045
0.1	11.801

Use this information to find a more accurate value of the integral I.

9. Use the Richardson extrapolation to compute $f'(1.0)$ and $f'(5.0)$ to five place accuracy with $f = (x + 0.5)^{-2}$. Use the central difference formula

$$f'(x) \approx \frac{f(x+h) - f(x-h)}{2h}$$

and take the initial step size, $h_o = 0.5$. Comment on the reason for the difference in the convergence rates for the two derivatives.

10. Use the Gauss quadrature to integrate:

$$I = \int_{-\infty}^{+\infty} e^{-x^2} \cos \alpha x \, dx$$

for $\alpha = 5$. The exact solution is $I = \sqrt{\pi} e^{-\alpha^2/4}$. The example worked out in the text corresponded to $\alpha = 1$. For the present case of $\alpha = 5$, discuss the number of function evaluations required to get the same level of accuracy as in the example.

11. Describe in detail an adaptive quadrature method that uses the trapezoidal rule as its basic integration scheme. Show in detail the error estimate.

12. Develop a quadrature method based on the cubic spline interpolation.

FURTHER READING

Abramowitz, M., and Stegun, I. *Handbook of Mathematical Functions with Formulas, Graphs, and Mathematical Tables.* Dover, 1972.

Dahlquist, G., and Björck, Å. *Numerical Methods*. Prentice-Hall, 1974, Chapter 7.

Ferziger, J. H. *Numerical Methods for Engineering Application*, Second Edition. Wiley, 1998, Chapter 3.

Forsythe, G. E., Malcolm, M. A., and Moler, C. B. *Computer Methods for Mathematical Computations*. Prentice-Hall, 1977, Chapter 5.

Press, W. H., Teukolsky, S. A., Vetterling, W. T., and Flannery, B. P. *Numerical Recipes: The Art of Scientific Computing*, Second Edition. Cambridge University Press, 1992, Chapter 4.

4

Numerical Solution of Ordinary Differential Equations

In this chapter we shall consider numerical solution of ordinary differential equations, ODEs. Here we will experience the real power of numerical analysis for engineering applications, as we will be able to tackle some real problems. We will consider both single and systems of differential equations. Since high-order ODEs can be converted to a system of first-order differential equations, our concentration will be on first-order ODEs. The extension to systems will be straightforward. We will consider all classes of ordinary differential equations: *initial*, *boundary* and *eigenvalue* problems. However, we will emphasize techniques for initial value problems because they are used extensively as the basis of methods for the other types of differential equations. The material in this chapter constitutes the core of this first course in numerical analysis; as we shall see in Chapter 5, numerical methods for partial differential equations are rooted in the methods for ODEs.

4.1 Initial Value Problems

Consider the first-order ordinary differential equation

$$\frac{dy}{dt} = f(y, t) \qquad y(0) = y_0. \tag{4.1}$$

We would like to find $y(t)$ for $0 < t \le t_f$. The aim of all numerical methods for solution of this differential equation is to obtain the solution at time $t_{n+1} = t_n + \Delta t$, given the solution for $0 \le t \le t_n$. This process, of course, continues; i.e., once $y_{n+1} = y(t_{n+1})$ is obtained, then y_{n+2} is calculated and so on until the final time, t_f.

We begin by considering the so-called Taylor series methods. Let's expand the solution at t_{n+1} about the solution at t_n

$$y_{n+1} = y_n + hy'_n + \frac{h^2}{2}y''_n + \frac{h^3}{6}y'''_n + \cdots \tag{4.2}$$

where $h = \Delta t$. From the differential equation (4.1), we have

$$y'_n = f(y_n, t_n)$$

which can be substituted in the second term in (4.2). We can, in principle, stop at this point, drop the higher order terms in (4.2), and get a second-order approximation to y_{n+1} using y_n. To get higher order approximations to y_{n+1}, we need to evaluate the higher order derivatives in (4.2) in terms of the known quantities at $t = t_n$. We will use the chain rule to obtain

$$y'' = \frac{dy'}{dt} = \frac{df}{dt} = \frac{\partial f}{\partial t} + \frac{\partial f}{\partial y}\frac{dy}{dt}$$
$$= f_t + f f_y$$

$$y''' = \frac{\partial}{\partial t}[f_t + f f_y] + \frac{\partial}{\partial y}[f_t + f f_y] f$$
$$= f_{tt} + 2 f f_{yt} + f_t f_y + f f_y^2 + f^2 f_{yy}.$$

Since f is a known function of y and t, all the above partial derivatives can, in principle, be computed. However, it is clear that the number of terms increases rapidly, and the method is not very practical for higher than third order.

The method based on the first two terms in the expansion is called the *Euler method*:

$$y_{n+1} = y_n + h f(y_n, t_n). \tag{4.3}$$

In using the Euler method, one simply starts from the initial condition, y_0, and marches forward using this formula to obtain $y_1, y_2, \ldots$. We will study the properties of this method extensively as it is a very simple method to analyze. From the Taylor series expansion it is apparent that the Euler method is second-order accurate for one time step. That is, if the exact solution is known at time step n, the numerical solution at time step $n + 1$ is second-order accurate. However, as with the quadrature formulas, in multi-step calculations, the errors accumulate, and the global error for advancing from the initial condition to the final time t_f is only *first-order* accurate.

Among the more accurate methods that we will discuss are the *Runge–Kutta* formulas. With explicit Runge–Kutta methods the solution at time step t_{n+1} is obtained in terms of y_n, $f(y_n, t_n)$, and $f(y, t)$ evaluated at the intermediate steps between t_n and $t_{n+1} = t_n + \Delta t$ (not including t_{n+1}). The higher accuracy is achieved because more information about f is provided due to the intermediate evaluations of f. This is in contrast to the Taylor series method where we provided more information about f through the higher derivatives of f at t_n.

Higher accuracy can also be obtained by providing information about f at times $t < t_n$. That is, the corresponding formulas involve $y_{n-1}, y_{n-2}, \ldots$, and $f_{n-1}, f_{n-2}, \ldots$. These methods are called *multi-step* methods.

We will also distinguish between *explicit* and *implicit* methods. The preceding methods were all explicit. The formulas that involve $f(y, t)$ evaluated at y_{n+1}, t_{n+1} belong to the class of implicit methods. Since f may be a non-linear function of y, to obtain the solution at each time step, implicit methods usually require solution of non-linear algebraic equations. Although the computational cost per time step is higher, implicit methods offer the advantage of numerical stability, which we shall discuss next.

4.2 Numerical Stability

So far, in the previous chapters, we have been concerned only with the accuracy of numerical methods and the work required to implement them. In this section the concept of numerical stability in numerical analysis is introduced, which is a more critical property of numerical methods for solving differential equations. It is quite possible for the numerical solution to a differential equation to grow unbounded even though its exact solution is well behaved. Of course, there are cases for which the exact solution grows unbounded, but for our discussion of stability, we shall concentrate only on the cases in which the exact solution is bounded. Given a differential equation

$$y' = f(y, t) \tag{4.1}$$

and a numerical method, in stability analysis we seek the conditions in terms of the parameters of the numerical method (mainly the step size h) for which the numerical solution remains bounded. In this context we have three classes of numerical methods:

Stable numerical scheme: Numerical solution does not grow unbounded (blow up) with any choice of parameters such as the step size. We will have to see what the cost is for such robustness.

Unstable numerical scheme: Numerical solution blows up with any choice of parameters. Clearly, no matter how accurate they may be, such numerical schemes would not be useful.

Conditionally stable: With certain choices of parameters the numerical solution remains bounded. Hopefully, the cost of the calculation does not become prohibitively large.

We would apply the so-called *stability analysis* to a numerical method to determine its stability properties, i.e., to determine to which of the above categories the method belongs. The analysis is performed for a simpler equation than (4.1), which hopefully retains some of the features of the general equation. Consider the two-dimensional Taylor series expansion of $f(y, t)$:

$$f(y,t) = f(y_0,t_0) + (t-t_0)\frac{\partial f}{\partial t}(y_0,t_0) + (y-y_0)\frac{\partial f}{\partial y}(y_0,t_0)$$

$$+\frac{1}{2!}\left[(t-t_0)^2\frac{\partial^2 f}{\partial t^2} + 2(t-t_0)(y-y_0)\frac{\partial^2 f}{\partial t \partial y} + (y-y_0)^2\frac{\partial^2 f}{\partial y^2}\right] + \cdots.$$

Collecting only the linear terms and substituting in (4.1), we formally get

$$y' = \lambda y + \alpha_1 + \alpha_2 t + \cdots \tag{4.4}$$

where $\lambda, \alpha_1, \alpha_2$ are constants. For example,

$$\lambda = \frac{\partial f}{\partial y}(y_0, t_0).$$

Discarding the non-linear terms (those involving higher powers of $(y - y_0)$, $(t - t_0)$ or their product) on the right-hand side of (4.4) yields the linearization of (4.1) about (y_0, t_0). For convenience and feasibility of analytical treatment, stability analysis is usually performed on the *model problem*, consisting of only the first term on the right-hand side of (4.4),

$$y' = \lambda y, \tag{4.5}$$

instead of the general problem (4.1). Here, λ is a constant. It turns out that the inhomogeneous terms in the linearized equation (4.4) do not significantly affect the results of the stability analysis. Note that the model equation has an exponential solution, which is the most dangerous part of the full solution of (4.1).

In our treatment of (4.5), we will allow λ to be complex

$$\lambda = \lambda_R + i\lambda_I$$

with the real part $\lambda_R \leq 0$ to ensure that the solution does not grow with t. This generalization will allow us to readily apply the results of our analysis to systems of ordinary differential equations and partial differential equations. To illustrate this point, consider the second-order differential equation

$$y'' + \omega^2 y = 0.$$

The exact solution is sinusoidal

$$y = c_1 \cos \omega t + c_2 \sin \omega t.$$

We can convert this second-order equation to two first-order equations

$$\begin{bmatrix} y_1 \\ y_2 \end{bmatrix}' = \begin{bmatrix} 0 & 1 \\ -\omega^2 & 0 \end{bmatrix}\begin{bmatrix} y_1 \\ y_2 \end{bmatrix}.$$

The eigenvalues of the 2×2 matrix A,

$$A = \begin{bmatrix} 0 & 1 \\ -\omega^2 & 0 \end{bmatrix},$$

are $\lambda = \pm i\omega$. Diagonalizing A with the matrix of its eigenvectors S,

$$A = S^{-1}\Lambda S,$$

leads to the uncoupled set of equations

$$z' = \Lambda z,$$

where

$$z = S\begin{pmatrix} y_1 \\ y_2 \end{pmatrix}$$

and Λ is the diagonal matrix with eigenvalues of A on the diagonal. The differential equations for the components of z are

$$z_1' = i\omega z_1 \qquad z_2' = -i\omega z_2.$$

This simple example illustrates that higher order linear differential equations or systems of first-order linear differential equations can reduce to uncoupled ordinary differential equations of the form of (4.5) with complex coefficients. The imaginary part of the coefficient results in oscillatory solutions of the form $e^{\pm i\omega t}$, and the real part dictates whether the solution grows or decays. For our stability analysis *we will be concerned only with cases where λ has a zero or negative real part.*

4.3 Stability Analysis for the Euler Method

Applying the Euler method (4.3),

$$y_{n+1} = y_n + hf(y_n, t_n),$$

to the model problem (4.5) leads to

$$\begin{aligned} y_{n+1} &= y_n + \lambda h y_n \\ &= y_n(1 + \lambda h). \end{aligned}$$

Thus, the solution at time step n can be written as

$$y_n = y_0(1 + \lambda h)^n. \tag{4.6}$$

For complex λ, we have

$$y_n = y_0(1 + \lambda_R h + i\lambda_I h)^n = y_0\sigma^n,$$

where $\sigma = (1 + \lambda_R h + i\lambda_I h)$ is called the amplification factor. The numerical solution is stable (i.e., remains bounded as n becomes large) if

$$|\sigma| \leq 1. \tag{4.7}$$

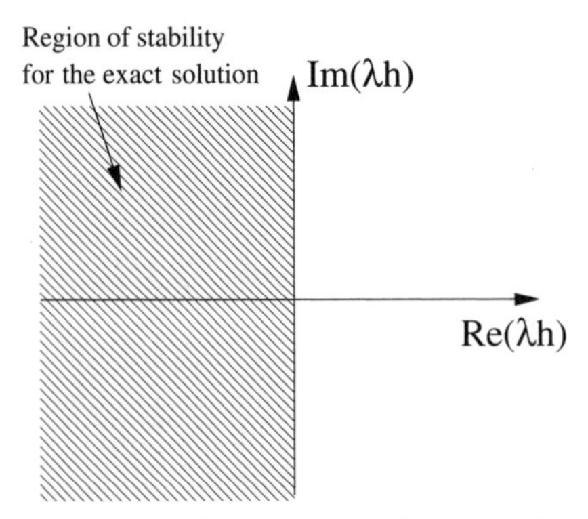

Figure 4.1 Stability diagram for the exact solution in the $\lambda_R h - \lambda_I h$ plane.

Note that for $\lambda_R \leq 0$ (which is the only case we consider) the exact solution, $y_0 e^{\lambda t}$, decays. That is, in the $(\lambda_R h - \lambda_I h)$ plane, the region of stability of the *exact solution* is the left-hand plane as illustrated in Figure 4.1.

However, only a portion of this plane is the region of stability for the Euler method. This portion is inside the circle

$$|\sigma|^2 = (1 + \lambda_R h)^2 + \lambda_I^2 h^2 = 1. \tag{4.8}$$

For any value of λh in the left-hand plane and outside this circle the numerical solution blows up while the exact solution decays (see Figure 4.2). Thus, the Euler method is *conditionally stable*. To have a stable numerical solution, we must reduce the step size h so that λh falls within the circle. If λ is *real* (and negative), then the maximum step size for stability is $2/|\lambda|$. That is, to get a stable solution, we must limit the step size to

$$h \leq \frac{2}{|\lambda|}. \tag{4.9}$$

Note that for real (and negative) λ, (4.7) is enforced for λh as low as -2. The main consequence of this limitation on h is that it would require more time steps, and hence more work, to reach the final time of integration, t_f. The circle (4.8)

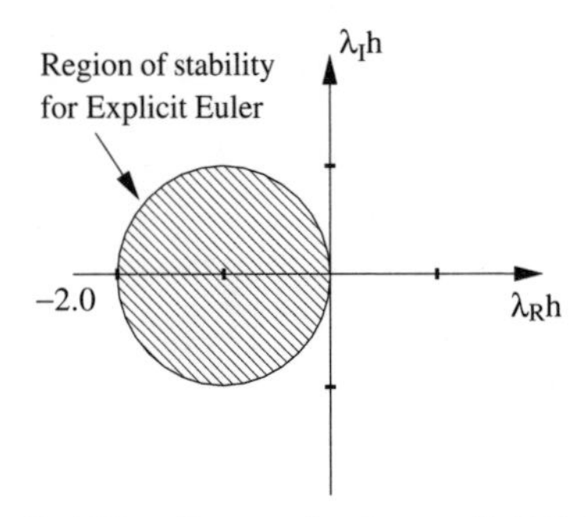

Figure 4.2 Stability diagram for the explicit Euler method.

is only tangent to the imaginary axis. Therefore, the Euler method is always unstable (irrespective of the step size) for *purely imaginary* λ. If λ is real and the numerical solution is unstable, then we must have

$$|1 + \lambda h| > 1,$$

which means that $(1 + \lambda h)$ is negative with magnitude greater than 1. Since

$$y_n = (1 + \lambda h)^n y_0,$$

the numerical solution exhibits oscillations with change of sign at every time step. This oscillatory behavior of the numerical solution is usually a good indication of numerical instability.

EXAMPLE 4.1 Explicit Euler

We will solve the following ODE using the Euler method:

$$y' + 0.5y = 0$$
$$y(0) = 1 \qquad 0 \le t \le 20.$$

Here λ is real and negative. The stability analysis of this section indicates that the Euler method should be stable for $h \le 4$. The solution is advanced by

$$y_{n+1} = y_n - 0.5hy_n$$

and the results for stable $(h = 1.0)$ and unstable $(h = 4.2)$ solutions are presented in Figure 4.3. We see that the solution with $h = 4.2$ is indeed unstable. Also note the oscillatory behavior of the solution before blow-up.

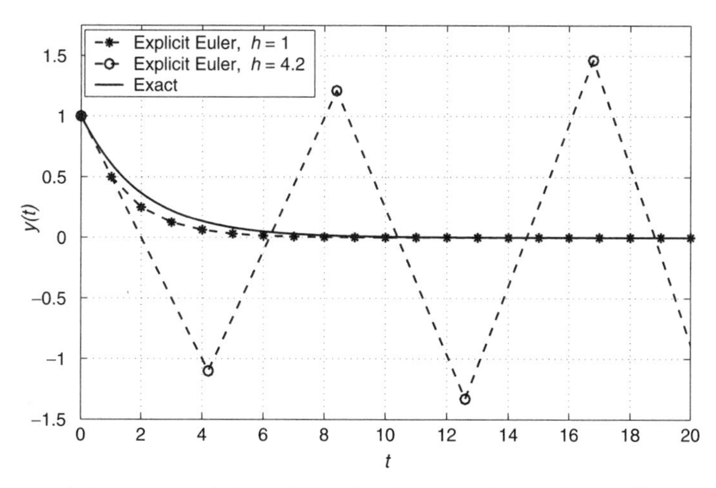

Figure 4.3 Numerical solution of the ODE in Example 1 using the Euler method.

4.4 Implicit or Backward Euler

The implicit Euler scheme is given by the following formula:

$$y_{n+1} = y_n + hf(y_{n+1}, t_{n+1}). \tag{4.10}$$

Note that in contrast to the explicit Euler, the implicit Euler does not allow us to easily obtain the solution at the next time step. If f is non-linear, we must solve a non-linear algebraic equation at each time step to obtain y_{n+1}, which usually requires an iterative algorithm. Therefore, the computational cost per time step for this scheme is, apparently, much higher than that for the explicit Euler. However, as we shall see below, the implicit Euler method has a much better stability property. Moreover, Section 4.7 will show that at each step, the requirement for an iterative algorithm may be avoided by the linearization technique.

Applying the backward Euler scheme to the model equation (4.5), we obtain

$$y_{n+1} = y_n + \lambda h y_{n+1}.$$

Solving for y_{n+1} produces

$$y_{n+1} = (1 - \lambda h)^{-1} y_n$$

or

$$y_n = \sigma^n y_0,$$

where

$$\sigma = \frac{1}{1 - \lambda h}.$$

Considering complex λ, we have

$$\sigma = \frac{1}{(1 - \lambda_R h) - i\lambda_I h}.$$

The denominator is a complex number and can be written as the product of its modulus and phase factor,

$$\sigma = \frac{1}{Ae^{i\theta}},$$

where

$$A = \sqrt{(1 - \lambda_R h)^2 + \lambda_I^2 h^2}, \qquad \theta = -\tan^{-1}\frac{\lambda_I h}{1 - \lambda_R h}.$$

For stability, the modulus of σ must be less than or equal to 1; i.e.,

$$|\sigma| = \frac{|e^{-i\theta}|}{A} = \frac{1}{A} \leq 1.$$

This is always true because λ_R is negative and hence $A > 1$. Thus, the backward Euler scheme is unconditionally stable. Unconditional stability is the usual characteristic of implicit methods. However, the price is higher computational cost per time step for having to solve a non-linear equation.

It should be pointed out that one can construct conditionally stable implicit methods. Obviously, such methods are not very popular because of the higher cost per step without the benefit of unconditional stability. Also note that numerical stability does not necessarily imply accuracy. A method can be stable but inaccurate. From the stability point of view, our objective is to use the maximum step size h to reach the final destination at time $t = t_f$. Large time steps translate to a lower number of function evaluations and lower computational cost. Large time steps may not be optimum for acceptable accuracy, but are strived for from the stability point of view.

EXAMPLE 4.2 Implicit (Backward) Euler

We now solve the ODE of Example 1 using the implicit Euler method. The stability analysis for the implicit Euler indicated that the numerical solution should be unconditionally stable. The solution is advanced by

$$y_{n+1} = \frac{y_n}{1 + 0.5h}$$

and the results for $h = 1.0$ and $h = 4.2$ are presented in Figure 4.4. Both solutions are now seen to be stable, as expected. The solution with $h = 1.0$ is more accurate. Note that the usual difficulty in obtaining the solution at each time step inherent with implicit methods is not encountered here because the differential equation in this example is linear.

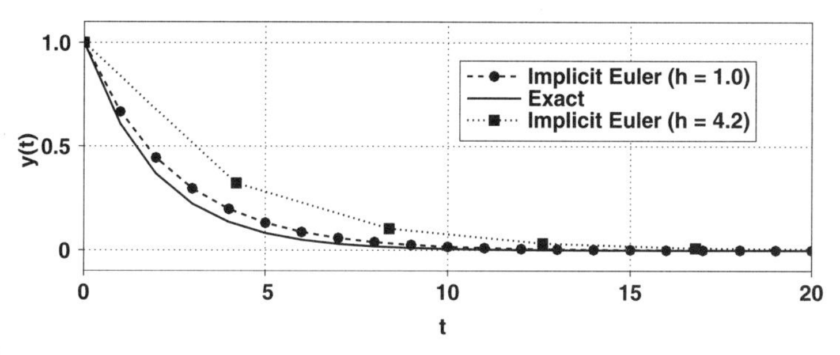

Figure 4.4 Numerical solution of the ODE in Example 2 using the implicit Euler method.

4.5 Numerical Accuracy Revisited

We have shown that the numerical solution to the model problem

$$y' = \lambda y \tag{4.5}$$

is of the form

$$y_n = y_0\sigma^n. \tag{4.11}$$

The exact solution is

$$y(t) = y_0 e^{\lambda t} = y_0 e^{\lambda nh} = y_0(e^{\lambda h})^n. \tag{4.12}$$

In analogy with the modified wavenumber approach of Chapter 2, one can often determine the order of accuracy of a method by comparing the numerical and exact solutions for a model problem, i.e., (4.11) and (4.12). That is, we compare the amplification factor σ with

$$e^{\lambda h} = 1 + \lambda h + \frac{\lambda^2 h^2}{2} + \frac{\lambda^3 h^3}{6} + \cdots.$$

For example, the amplification factor of the explicit Euler is

$$\sigma = 1 + \lambda h,$$

and the amplification factor for the backward Euler is

$$\sigma = \frac{1}{(1-\lambda h)} = 1 + \lambda h + \lambda^2 h^2 + \lambda^3 h^3 + \cdots.$$

Thus, both methods are able to reproduce only up to the λh term in the exponential expansion. Each method is second-order accurate for one time step, but globally first order. From now on, we will call a method αth order if its amplification factor matches all the terms up to and including the $\lambda^\alpha h^\alpha/\alpha!$ term in the exponential expansion. The order of accuracy derived in this manner from the linear analysis (i.e., from application to (4.5)) should be viewed as the upper limit on the order of accuracy. A method may have a lower order of accuracy for non-linear equations.

Often the order of accuracy by itself is not very informative. In particular, in problems with oscillatory solutions, one is interested in the phase and amplitude errors separately. To understand this type of error analysis, we will consider the model equation with pure imaginary λ:

$$y' = i\omega y \qquad y(0) = 1.$$

The exact solution is $e^{i\omega t}$, which is oscillatory. The frequency of oscillations is ω and its amplitude is 1. The numerical solution with the explicit Euler is

$$y_n = \sigma^n y_0$$

where $\sigma = 1 + i\omega h$. It is clear that the amplitude of the numerical solution,

$$|\sigma| = \sqrt{1 + w^2 h^2}$$

is greater than 1, which reconfirms that the Euler method is unstable for purely imaginary λ. σ is a complex number and can be written as

$$\sigma = |\sigma| e^{i\theta},$$

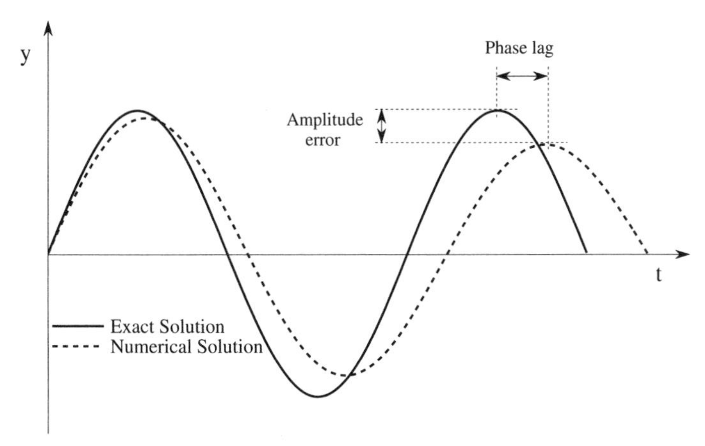

Figure 4.5 A schematic showing the amplitude and phase errors in the numerical solution.

where

$$\theta = \tan^{-1} \omega h = \tan^{-1} \frac{\text{Im}(\sigma)}{\text{Re}(\sigma)}.$$

A measure of the phase error (PE) (see Figure 4.5) is obtained from comparison with the phase of the exact solution

$$\text{PE} = \omega h - \theta = \omega h - \tan^{-1} \omega h.$$

Using the power series for $\tan^{-1}$,

$$\tan^{-1} \omega h = \omega h - \frac{(\omega h)^3}{3} + \frac{(\omega h)^5}{5} - \frac{(\omega h)^7}{7} + \cdots$$

we have

$$\text{PE} = \frac{(\omega h)^3}{3} + \cdots, \tag{4.13}$$

which corresponds to a phase lag. This is the phase error encountered at each step. The phase error after n time steps is nPE.

4.6 Trapezoidal Method

The formal solution to the differential equation (4.1) with the condition $y(t_n) = y_n$ is

$$y(t) = y_n + \int_{t_n}^{t} f(y, t')\, dt'.$$

At $t = t_{n+1}$

$$y_{n+1} = y_n + \int_{t_n}^{t_{n+1}} f(y, t')\, dt'.$$

Approximating the integral with the trapezoidal method leads to

$$y_{n+1} = y_n + \frac{h}{2}[f(y_{n+1}, t_{n+1}) + f(y_n, t_n)]. \tag{4.14}$$

This is the trapezoidal method for the solution of ordinary differential equations. When applied to certain partial differential equations it is often called the Crank–Nicolson method. Clearly the trapezoidal method is an implicit scheme.

Applying the trapezoidal method to the model equation yields

$$y_{n+1} - y_n = \frac{h}{2}[\lambda y_{n+1} + \lambda y_n]$$

or

$$y_{n+1} = \frac{1 + \frac{\lambda h}{2}}{1 - \frac{\lambda h}{2}} y_n.$$

Expanding the amplification factor σ leads to

$$\sigma = \frac{1 + \frac{\lambda h}{2}}{1 - \frac{\lambda h}{2}} = 1 + \lambda h + \frac{\lambda^2 h^2}{2} + \frac{\lambda^3 h^3}{4} + \cdots$$

which indicates that the method is *second-order* accurate. The extra accuracy is obtained at virtually no extra cost over the backward Euler method.

Now, we will examine the stability properties of the trapezoidal method by computing the modulus of σ for complex $\lambda = \lambda_R + i\lambda_I$. The amplification factor becomes

$$\sigma = \frac{1 + \frac{\lambda_R h}{2} + i\frac{\lambda_I h}{2}}{1 - \frac{\lambda_R h}{2} - i\frac{\lambda_I h}{2}}.$$

Both the numerator and denominator are complex and can be written as $Ae^{i\theta}$ and $Be^{i\alpha}$, respectively, where

$$A = \sqrt{\left(1 + \frac{\lambda_R h}{2}\right)^2 + \frac{\lambda_I^2 h^2}{4}}$$

and

$$B = \sqrt{\left(1 - \frac{\lambda_R h}{2}\right)^2 + \frac{\lambda_I^2 h^2}{4}}.$$

Thus,

$$\sigma = \frac{A}{B} e^{i(\theta - \alpha)}$$

or

$$|\sigma| = \frac{A}{B}.$$

Since we are only interested in cases where $\lambda_R < 0$, and for these cases $A < B$, it follows that

$$|\sigma| < 1.$$

Thus, the trapezoidal method is unconditionally stable, which is expected since it is an *implicit* method. Note, however, that for real and negative λ,

$$\lim_{h\to\infty} \sigma = -1,$$

which implies that for large time steps, the numerical solution $\sigma^n y_0$ oscillates between y_0 and $-y_0$ from one time step to the next, but the solution will not blow up.

Let us examine the accuracy of the trapezoidal method for oscillatory solutions, $\lambda = i\omega$. In this case ($\lambda_R = 0$), $A = B$, and

$$|\sigma| = 1.$$

Thus, there is *no amplitude error* associated with the trapezoidal method. Since

$$\sigma = e^{2i\theta} \qquad \theta = \tan^{-1}\left(\frac{\omega h}{2}\right),$$

the phase error is given by

$$\text{PE} = \omega h - 2\tan^{-1}\frac{\omega h}{2} = \omega h - 2\left[\frac{\omega h}{2} - \frac{(\omega h)^3}{24} + \cdots\right] = \frac{(\omega h)^3}{12} + \cdots$$

which is about four times better than that for the explicit Euler but of the same order of accuracy.

EXAMPLE 4.3 A Second-Order Equation

We now consider the second-order equation

$$y'' + \omega^2 y = 0 \quad t > 0$$
$$y(0) = y_o \quad y'(0) = 0,$$

and investigate the numerical solutions by the explicit Euler, implicit Euler, and trapezoidal methods. In Section 4.2 it was demonstrated how this equation could be reduced to a coupled pair of first-order equations:

$$y_1' = y_2 \qquad y_2' = -\omega^2 y_1.$$

In matrix form we have

$$\begin{bmatrix} y_1 \\ y_2 \end{bmatrix}' = \begin{bmatrix} 0 & 1 \\ -\omega^2 & 0 \end{bmatrix} \begin{bmatrix} y_1 \\ y_2 \end{bmatrix}.$$

These equations were then decoupled, giving

$$z_1' = i\omega z_1 \qquad z_2' = -i\omega z_2.$$

The stability of the numerical solution depends upon the eigenvalues $i\omega$ and $-i\omega$ that decouple the system. We see that here the eigenvalues are

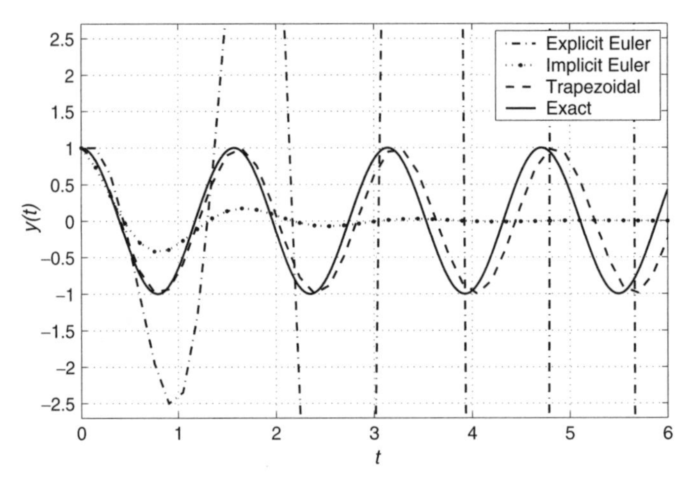

Figure 4.6 Numerical solution of the ODE in Example 3.

imaginary and therefore predict the Euler solution to be unconditionally unstable. We have also seen that both backward Euler and trapezoidal methods are unconditionally stable. We will show this to be the case by numerical simulation of the equations. Solution advancement proceeds as follows. For explicit Euler:

$$\begin{bmatrix} y_1 \\ y_2 \end{bmatrix}_{n+1} = \begin{bmatrix} 1 & h \\ -\omega^2 h & 1 \end{bmatrix} \begin{bmatrix} y_1 \\ y_2 \end{bmatrix}_n .$$

For implicit Euler:

$$\begin{bmatrix} 1 & -h \\ \omega^2 h & 1 \end{bmatrix} \begin{bmatrix} y_1 \\ y_2 \end{bmatrix}_{n+1} = \begin{bmatrix} y_1 \\ y_2 \end{bmatrix}_n .$$

For trapezoidal:

$$\begin{bmatrix} 1 & -\frac{h}{2} \\ \omega^2 \frac{h}{2} & 1 \end{bmatrix} \begin{bmatrix} y_1 \\ y_2 \end{bmatrix}_{n+1} = \begin{bmatrix} 1 & \frac{h}{2} \\ -\omega^2 \frac{h}{2} & 1 \end{bmatrix} \begin{bmatrix} y_1 \\ y_2 \end{bmatrix}_n .$$

Numerical results are plotted in Figure 4.6 for $y_o = 1$, $\omega = 4$, and time step $h = 0.15$.

We see that the explicit Euler rapidly blows up as expected. The implicit Euler is stable, but decays very rapidly. The trapezoidal method performs the best and has zero amplitude error as predicted in the analysis of Section 4.6; however, its phase error is evident and is increasing as the solution proceeds.

Although the numerical methods used in the previous example were introduced in the context of a single differential equation, their application to a system was a straightforward generalization of the corresponding single equation formulas. It is also important to emphasize that the decoupling of the equations using eigenvalues and eigenvectors was performed solely for the purpose of stability analysis. The equations are never decoupled in actual numerical solutions.

4.7 Linearization for Implicit Methods

As pointed out in Section 4.4, the difficulty with implicit methods is that, in general, at each time step, they requires solving a non-linear algebraic equation, which often requires an iterative solution procedure such as the Newton–Raphson method. For non-linear initial value problems, iteration can be avoided by the *linearization technique.* Consider the ordinary differential equation:

$$y' = f(y, t). \tag{4.1}$$

Applying the trapezoidal method to this equation yields

$$y_{n+1} = y_n + \frac{h}{2}[f(y_{n+1}, t_{n+1}) + f(y_n, t_n)] + O(h^3). \tag{4.15}$$

To solve for y_{n+1} would require solving a non-linear algebraic equation, and non-linear equations are usually solved by iterative methods. However, by realizing that (4.15) is already an approximate equation (to $O(h^3)$), it would not make sense to find its solution exactly or to within round-off error. Therefore, we will attempt to solve the non-linear equation (4.15) to $O(h^3)$, which, hopefully, will not require iterations.

Consider the Taylor series expansion of $f(y_{n+1}, t_{n+1})$:

$$\begin{aligned} f(y_{n+1}, t_{n+1}) = f(y_n, t_{n+1}) &+ (y_{n+1} - y_n)\frac{\partial f}{\partial y}\bigg|_{(y_n, t_{n+1})} \\ &+ \frac{1}{2}(y_{n+1} - y_n)^2 \frac{\partial^2 f}{\partial y^2}\bigg|_{(y_n, t_{n+1})} + \cdots . \end{aligned} \tag{4.16}$$

But from Taylor series expansion for y we have

$$y_{n+1} - y_n = O(h).$$

Therefore, replacing $f(y_{n+1}, t_{n+1})$ in (4.15) with the first two terms in its Taylor series expansion does not alter the order of accuracy of (4.15), which (for one step) is $O(h^3)$. Making this substitution results in

$$y_{n+1} = y_n + \frac{h}{2}\left[f(y_n, t_{n+1}) + (y_{n+1} - y_n)\frac{\partial f}{\partial y}\bigg|_{(y_n, t_{n+1})} + f(y_n, t_n)\right] + O(h^3). \tag{4.17}$$

Rearranging and solving for y_{n+1}, yields

$$y_{n+1} = y_n + \frac{h}{2}\frac{f(y_n, t_{n+1}) + f(y_n, t_n)}{1 - \frac{h}{2}\frac{\partial f}{\partial y}\Big|_{(y_n, t_{n+1})}}. \tag{4.18}$$

Thus, the solution can proceed without iteration while retaining the global second-order accuracy. Clearly, as far as the *linear* stability analysis is concerned, the linearized scheme is also unconditionally stable. However, one should caution that in practice, linearization may lead to some loss of total stability for non-linear f.

EXAMPLE 4.4 Linearization

We consider the non-linear ordinary differential equation

$$y' + y(1 - y) = 0 \qquad y(0) = \frac{1}{2}$$

and its numerical solution by the trapezoidal method:

$$y_{n+1} = y_n + \frac{h}{2}[y_{n+1}(y_{n+1} - 1) + y_n(y_n - 1)].$$

This, of course, is a non-linear algebraic equation for y_{n+1}. Using the linearization method developed in this section, where f is now $y(y - 1)$, we arrive at the following linearized trapezoidal method:

$$y_{n+1} = y_n + \frac{hy_n(y_n - 1)}{1 - h\left(y_n - \frac{1}{2}\right)}.$$

Since the non-linearity is quadratic, we may also solve the resulting non-linear algebraic equation directly and compare the direct implicit solution with the linearized solution. The direct implicit solution is given by

$$y_{n+1} = \frac{\left(\frac{2}{h} + 1\right) - \sqrt{\left(\frac{2}{h} + 1\right)^2 - 4\left(\frac{2}{h}y_n + y_n(y_n - 1)\right)}}{2}.$$

These equations were advanced from time $t = 0$ to $t = 1$. The error in the solution at $t = 1$ is plotted in Figure 4.7 versus the number of steps taken. The slopes for both the trapezoidal and linearized trapezoidal methods clearly show a second-order dependence upon number of steps, demonstrating that second-order accuracy is maintained with linearization. The directly solved trapezoidal method is slightly more accurate, but this is a problem-specific phenomenon (for example, the linearized trapezoidal solution for $y' + y^2 = 0$ yields the *exact* solution for any h while the accuracy of the direct implicit solution is dependent on h).

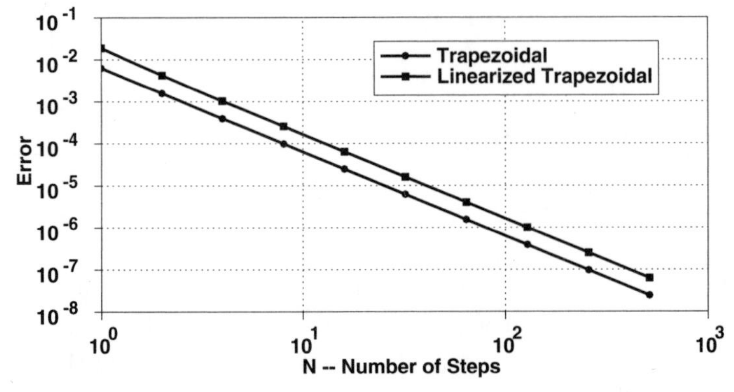

Figure 4.7 Error in the solution of the ODE in Example 4.

4.8 Runge–Kutta Methods

We noted in the Taylor series method, in Section 4.1, that the order of accuracy of a method increases by including more terms in the expansion. The additional terms involve various partial derivatives of $f(y, t)$, which provide additional information on f at $t = t_n$. Note that the analytical form of f is not transparent to a time-stepping procedure, only numerical data at one or more steps are. There are different methods of providing additional information about f. Runge–Kutta (RK) methods introduce points between t_n and t_{n+1} and evaluate f at these intermediate points. The additional function evaluations, of course, result in higher cost per time step; but the accuracy is increased, and as it turns out, better stability properties are also obtained.

We begin by describing the general form of (two stage) *second-order* Runge–Kutta formulas for solving

$$y' = f(y, t). \tag{4.1}$$

The solution at time step t_{n+1} is obtained from

$$y_{n+1} = y_n + \gamma_1 k_1 + \gamma_2 k_2, \tag{4.19}$$

where the functions k_1 and k_2 are defined sequentially

$$k_1 = hf(y_n, t_n) \tag{4.20}$$

$$k_2 = hf(y_n + \beta k_1, t_n + \alpha h), \tag{4.21}$$

and $\alpha, \beta, \gamma_1, \gamma_2$ are constants to be determined. These constants are determined to ensure the highest order of accuracy for the method. To establish the order of accuracy, consider the Taylor series expansion of $y(t_{n+1})$ from Section 4.1:

$$y_{n+1} = y_n + hy'_n + \frac{h^2}{2} y''_n + \cdots .$$

But

$$y'_n = f(y_n, t_n),$$

and using the chain rule, we have already obtained

$$y'' = f_t + f f_y,$$

where f_t and f_y are the partial derivatives of f with respect to t and y respectively. Thus,

$$y_{n+1} = y_n + hf(y_n, t_n) + \frac{h^2}{2}(f_{t_n} + f_n f_{y_n}) + \cdots . \tag{4.22}$$

To establish the order of accuracy of the Runge–Kutta method as given by (4.19), we must compare its estimate for y_{n+1} to that of the Taylor series formula (4.22). For this comparison to be useful, we must convert the various terms in these expressions into common forms. Two-dimensional Taylor series expansion of k_2 (4.21) leads to

$$k_2 = h[f(y_n, t_n) + \beta k_1 f_{y_n} + \alpha h f_{t_n} + O(h^2)].$$

Noting that $k_1 = hf(y_n, t_n)$ and substituting in (4.19) yields

$$y_{n+1} = y_n + (\gamma_1 + \gamma_2)hf_n + \gamma_2 \beta h^2 f_n f_{y_n} + \gamma_2 \alpha h^2 f_{t_n} + \cdots. \tag{4.23}$$

Comparison of (4.22) and (4.23) and matching coefficients of similar terms leads to

$$\gamma_1 + \gamma_2 = 1$$

$$\gamma_2 \alpha = \frac{1}{2}$$

$$\gamma_2 \beta = \frac{1}{2}.$$

These are three non-linear equations for the four unknowns. Using α as a free parameter, we have

$$\gamma_2 = \frac{1}{2\alpha} \qquad \beta = \alpha \qquad \gamma_1 = 1 - \frac{1}{2\alpha}.$$

With three out of the four constants chosen, we have a one-parameter family of second-order Runge–Kutta formulas:

$$k_1 = hf(y_n, t_n) \tag{4.24a}$$

$$k_2 = hf(y_n + \alpha k_1, t_n + \alpha h) \tag{4.24b}$$

$$y_{n+1} = y_n + \left(1 - \frac{1}{2\alpha}\right) k_1 + \frac{1}{2\alpha} k_2. \tag{4.24c}$$

Thus, we have a second-order Runge–Kutta formula for each value of α chosen. The choice $\alpha = 1/2$ is made frequently. In actual computations, one calculates k_1 using (4.24a); this value is then used to compute k_2 using (4.24b) followed by the calculation of y_{n+1} using (4.24c).

Runge–Kutta formulas are often presented in a different but equivalent form. For example, the popular form of the second-order Runge–Kutta formula ($\alpha = 1/2$) is presented in the following (predictor–corrector) format:

$$y^*_{n+1/2} = y_n + \frac{h}{2} f(y_n, t_n) \tag{4.25a}$$

$$y_{n+1} = y_n + hf(y^*_{n+1/2}, t_{n+1/2}). \tag{4.25b}$$

Here, one calculates the *predicted* value in (4.25a) which is then used in (4.25b) to obtain the *corrected* value, y_{n+1}.

Now, let's use linear analysis to gain insight into the stability and accuracy of the second order Runge–Kutta method discussed above. Applying the Runge–Kutta method in (4.24) to the model equation $y' = \lambda y$ results in

$$k_1 = \lambda h y_n$$

$$k_2 = h(\lambda y_n + \alpha\lambda^2 h y_n) = \lambda h(1 + \alpha h\lambda) y_n$$

$$y_{n+1} = y_n + \left(1 - \frac{1}{2\alpha}\right)\lambda h y_n + \frac{1}{2\alpha}\lambda h(1 + \alpha\lambda h) y_n$$

$$= y_n\left(1 + \lambda h + \frac{\lambda^2 h^2}{2}\right). \tag{4.26}$$

Thus, we have a confirmation that the method is second-order accurate. For stability, we must have $|\sigma| \leq 1$, where

$$\sigma = \left(1 + \lambda h + \frac{\lambda^2 h^2}{2}\right). \tag{4.27}$$

A convenient way to obtain the stability *boundary*, i.e., $|\sigma| = 1$, of the method is to set

$$\sigma = \left(1 + \lambda h + \frac{\lambda^2 h^2}{2}\right) = e^{i\theta}$$

and find the complex roots λh of this polynomial for different values of θ. Recall that $|e^{i\theta}| = 1$ for all values of θ. The resulting stability region is shown in Figure 4.8. On the real axis the stability boundary is the same as that of explicit Euler ($|\lambda_R h| \leq 2$); however, there is significant improvement for complex λ. The method is also unstable for purely imaginary λ. In this case, substituting $\lambda = i\omega$ into (4.27) results in

$$|\sigma| = \sqrt{1 + \frac{\omega^4 h^4}{4}} > 1, \tag{4.28}$$

i.e., the method is unconditionally unstable for purely imaginary λ. However, note that for small values of ωh, this method is less unstable than explicit Euler.

EXAMPLE 4.5 Amplification Factor

Let's consider numerical solution of

$$y' = i\omega y \qquad y(0) = 1$$

using the explicit Euler method and a second-order Runge–Kutta scheme. Suppose the differential equation is integrated for 100 time steps with $\omega h = 0.2$; that is, the integration time is from $t = 0$ to $t = 20/\omega$. Each numerical solution after 100 time steps can be written as

$$y = \sigma^{100} y_0,$$

where σ is the corresponding amplification factor for each method. For the Euler scheme, $|\sigma| = \sqrt{1+\omega^2 h^2} = 1.0198$, and for the RK method, from (4.28), we have $|\sigma| = 1.0002$. Thus, after 100 time steps, for the RK method we have $y = 1.02$, i.e., there is only 2% amplitude error, whereas for the Euler method we have $y = 7.10$!

The phase error for the second-order RK scheme is easily calculated from the real and imaginary parts of σ for the case $\lambda = i\omega$:

$$\mathrm{PE} = \omega h - \tan^{-1}\left(\frac{\omega h}{1 - \frac{\omega^2 h^2}{2}}\right).$$

But

$$\tan^{-1}\left(\frac{\omega h}{1 - \frac{\omega^2 h^2}{2}}\right) = \omega h \left(1 + \frac{\omega^2 h^2}{2} + \frac{\omega^4 h^4}{4} + \cdots\right)$$
$$-\frac{1}{3}\left[\omega h \left(1 + \frac{\omega^2 h^2}{2} + \frac{\omega^4 h^4}{4} + \cdots\right)\right]^3 + \cdots = \omega h + \frac{\omega^3 h^3}{6} + \cdots.$$

Hence,

$$\mathrm{PE} = -\frac{\omega^3 h^3}{6} + \cdots, \tag{4.29}$$

which is only a factor of 2 better than Euler, but of opposite sign. Negative phase error corresponds to phase lead (see Example 6).

The most widely used Runge–Kutta method is the fourth-order formula. This is perhaps the most popular numerical scheme for initial value problems. The fourth-order formula can be presented in a typical RK format:

$$y_{n+1} = y_n + \frac{1}{6}k_1 + \frac{1}{3}(k_2 + k_3) + \frac{1}{6}k_4, \tag{4.30a}$$

where

$$k_1 = hf(y_n, t_n) \tag{4.30b}$$

$$k_2 = hf\left(y_n + \frac{1}{2}k_1, t_n + \frac{h}{2}\right) \tag{4.30c}$$

$$k_3 = hf\left(y_n + \frac{1}{2}k_2, t_n + \frac{h}{2}\right) \tag{4.30d}$$

$$k_4 = hf(y_n + k_3, t_n + h). \tag{4.30e}$$

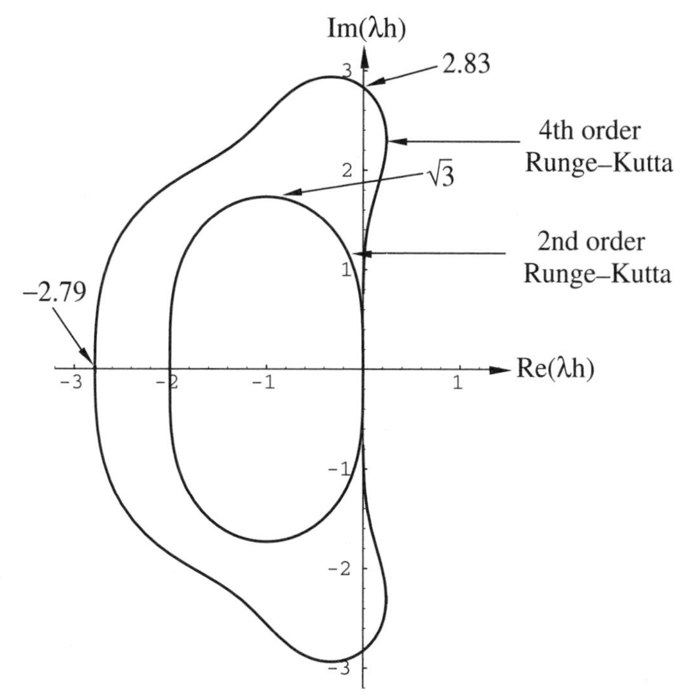

Figure 4.8 Stability diagrams for second- and fourth-order Runge–Kutta methods.

Note that four function evaluations are required at each time step. Applying the method to the model equation, $y' = \lambda y$, leads to

$$y_{n+1} = \left(1 + \lambda h + \frac{\lambda^2 h^2}{2} + \frac{\lambda^3 h^3}{6} + \frac{\lambda^4 h^4}{24}\right) y_n, \tag{4.31}$$

which confirms the fourth-order accuracy of the method. Again, the stability diagram is obtained by finding the roots of the following fourth-order polynomial with complex coefficients:

$$\lambda h + \frac{\lambda^2 h^2}{2} + \frac{\lambda^3 h^3}{6} + \frac{\lambda^4 h^4}{24} + 1 - e^{i\theta} = 0,$$

for different values of $0 \leq \theta \leq \pi$. This requires a root-finder for polynomials with complex coefficients. The resulting region of stability (Figure 4.8) shows a significant improvement over that obtained by the second-order Runge–Kutta. In particular, it has a large stability region on the imaginary axis. In fact there are two small stable regions corresponding to positive $Re(\lambda)$, where the exact solution actually grows; that is, the method is artificially stable for the parameters corresponding to these regions.

EXAMPLE 4.6 Runge–Kutta

We solve the problem of Example 3 using second- and fourth-order Runge–Kutta algorithms. The details for the second-order Runge–Kutta advancement are

$$y_1^{(n+1/2)*} = y_1^n + \frac{h}{2} y_2^n$$

$$y_2^{(n+1/2)*} = y_2^n - \frac{h}{2}\omega^2 y_1^n$$

$$y_1^{n+1} = y_1^n + h y_2^{(n+1/2)*}$$

$$y_2^{n+1} = y_2^n - h\omega^2 y_1^{(n+1/2)*}.$$

Fourth-order Runge–Kutta advancement proceeds similarly. Again numerical results are plotted in Figure 4.9 for $y_o = 1$, $\omega = 4$, and time step $h = 0.15$.

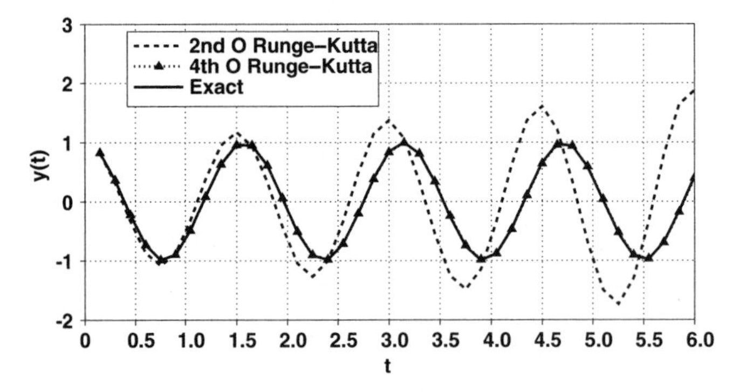

Figure 4.9 Numerical solution of the ODE in Example 3 using Runge–Kutta methods.

It can be seen that the second-order scheme is mildly unstable as predicted by the linear stability analysis. The fourth-order Runge–Kutta solution is stable as predicted and is highly accurate, showing to plotting accuracy, virtually no phase or amplitude errors.

The most expensive part of numerical solution of ordinary differential equations is the function evaluations. The number of steps (or the step size h) required to reach the final integration time t_f is therefore directly related to the cost of the computation. Hence, both the stability characteristics and the accuracy come into play in establishing the cost-effectiveness of a numerical method. The fourth-order Runge–Kutta scheme requires four function evaluations per time step. However, it also has superior stability as well as excellent accuracy properties. These characteristics, together with its ease of programming, have made the fourth-order RK one of the most popular schemes for the solution of ordinary and partial differential equations.

Finally, note that the order of accuracy of the second- and fourth-order Runge–Kutta formulas, discussed in this section, also corresponded to their respective number of function evaluations (stages). It turns out that this trend does not continue beyond fourth order. For example, a fifth-order Runge–Kutta formula requires six function evaluations.

4.9 Multi-Step Methods

The Runge–Kutta formulas obtained higher order accuracy through the use of several function evaluations. However, higher order accuracy can also be achieved by using data from prior to t_n; that is, if the solution and/or f at $t_{n-1}, t_{n-2}, \ldots$ are used. This is another way of providing additional information about f. Methods that use information from prior to step n are called multi-step schemes. The apparent price for the higher order of accuracy is the use of additional computer memory, which can be of concern for partial differential equations, as discussed in Chapter 5. Multi-step methods are *not* self-starting. Usually another method such as the explicit Euler is used to start the calculations for the first or the first few time steps.

A classical multi-step method is the *leapfrog method:*

$$y_{n+1} = y_{n-1} + 2hf(y_n, t_n) + O(h^3). \tag{4.32}$$

This method is derived by applying the second-order central difference formula for y' in (4.1). Thus, the leapfrog method is a second-order method. Starting with an initial condition y_0, a self-starting method like Euler is used to obtain y_1, and then leapfrog is used for steps two and higher. Applying leapfrog to the model equation, $y' = \lambda y$, leads to

$$y_{n+1} - y_{n-1} = 2\lambda h y_n.$$

This is a difference equation for y_n that cannot be solved as readily as the schemes discussed up to this point. To solve it, we assume a solution of the form

$$y_n = \sigma^n y_0.$$

Substitution in the difference equation leads to

$$\sigma^{n+1} - \sigma^{n-1} = 2h\lambda\sigma^n.$$

Dividing by σ^{n-1}, we will get a quadratic equation for σ

$$\sigma^2 - 2h\lambda\sigma - 1 = 0,$$

which can be solved to yield

$$\sigma_{1,2} = \lambda h \pm \sqrt{\lambda^2 h^2 + 1}.$$

Having more than one root is the key characteristic of multi-step methods. For comparison with the exponential solution to the model problem, we expand the roots in powers of λh

$$\sigma_1 = \lambda h + \sqrt{\lambda^2 h^2 + 1} = 1 + \lambda h + \frac{1}{2}\lambda^2 h^2 - \frac{1}{8}\lambda^4 h^4 + \cdots$$

$$\sigma_2 = \lambda h - \sqrt{\lambda^2 h^2 + 1} = -1 + \lambda h - \frac{1}{2}\lambda^2 h^2 + \frac{1}{8}\lambda^4 h^4 + \cdots.$$

The first root shows that the method is second-order accurate. The second root is spurious and often is a source of numerical problems. Note that even for $h = 0$, the spurious root is not equal to 1. It is also apparent that for λ real and negative, the spurious root has a magnitude greater than 1 which leads to instability.

Since the difference equation for y_n is linear, its general solution can be written as a linear combination of the roots, i.e.,

$$y_n = c_1\sigma_1^n + c_2\sigma_2^n. \tag{4.33}$$

That is, the solution is composed of contributions from both physical and spurious roots. The constants c_1 and c_2 are obtained from the starting conditions y_0 and y_1 by letting $n = 0$ and $n = 1$, respectively, in (4.33):

$$y_0 = c_1 + c_2 \qquad y_1 = c_1\sigma_1 + c_2\sigma_2.$$

Solving for c_1 and c_2 leads to

$$c_1 = \frac{y_1 - y_0\sigma_2}{\sigma_1 - \sigma_2} \qquad c_2 = \frac{\sigma_1 y_0 - y_1}{\sigma_1 - \sigma_2}.$$

Thus, *for the model problem*, if we choose $y_1 = \sigma_1 y_0$, the spurious root is completely suppressed. In general, we can expect the starting scheme to play a role in determining the level of contribution of the spurious root. Even if the spurious root is suppressed initially, round-off errors will restart it again. In the case of leapfrog, the spurious root leads to oscillations from one step to the next.

Application of leapfrog to the case where $\lambda = i\omega$ is pure imaginary leads to

$$\sigma_{1,2} = i\omega h \pm \sqrt{1 - \omega^2 h^2}.$$

If $|\omega h| \leq 1$, then

$$|\sigma_{1,2}| = 1.$$

In this case leapfrog has no amplitude error. This is the main reason for the use of leapfrog method. If $|\omega h| > 1$, then

$$|\sigma_{1,2}| = |\omega h \pm \sqrt{\omega^2 h^2 - 1}|$$

and the method is unstable.

Finally, we present the widely used second-order *Adams–Bashforth method*. This method can be easily derived by using the Taylor series expansion of y_{n+1}:

$$y_{n+1} = y_n + hy_n' + \frac{h^2}{2}y_n'' + \frac{h^3}{6}y_n''' + \cdots.$$

Substituting

$$y_n' = f(y_n, t_n),$$

and a first-order finite difference approximation for y_n''

$$y_n'' = \frac{f(y_n, t_n) - f(y_{n-1}, t_{n-1})}{h} + O(h)$$

leads to

$$y_{n+1} = y_n + \frac{3h}{2} f(y_n, t_n) - \frac{h}{2} f(y_{n-1}, t_{n-1}) + O(h^3). \tag{4.34}$$

Thus, the Adams–Bashforth method is second-order accurate *globally*. Applying the method to the model problem leads to the following second-order difference equation for y_n:

$$y_{n+1} - \left(1 + \frac{3\lambda h}{2}\right) y_n + \frac{\lambda h}{2} y_{n-1} = 0.$$

Once again assuming solutions of the form $y_n = \sigma^n$ results in a quadratic equation for σ with roots

$$\sigma_{1,2} = \frac{1}{2}\left[1 + \frac{3}{2}\lambda h \pm \sqrt{1 + \lambda h + \frac{9}{4}\lambda^2 h^2}\right].$$

Using the power series expansion for the square root

$$\sqrt{1 + \lambda h + \frac{9}{4}\lambda^2 h^2} = 1 + \frac{1}{2}\left(\lambda h + \frac{9}{4}\lambda^2 h^2\right) - \frac{1}{8}\left(\lambda h + \frac{9}{4}\lambda^2 h^2\right)^2 + \frac{3}{48}\left(\lambda h + \frac{9}{4}\lambda^2 h^2\right)^3 + \cdots,$$

we obtain

$$\sigma_1 = 1 + \lambda h + \frac{1}{2}\lambda^2 h^2 + O(h^3)$$

and

$$\sigma_2 = \frac{1}{2}\lambda h - \frac{1}{2}\lambda^2 h^2 + O(h^3).$$

The spurious root for the Adams–Bashforth method appears to be less dangerous. Observe that it approaches zero if $h \to 0$. The stability region of the Adams–Bashforth method is shown in Figure 4.10. It is oval-shaped in the $\lambda_R h - \lambda_I h$ plane. It crosses the real axis at -1, which is more limiting than the explicit Euler and second-order Runge–Kutta methods. It is also only tangent to the imaginary axis. Thus, strictly speaking, it is unstable for pure imaginary λ,

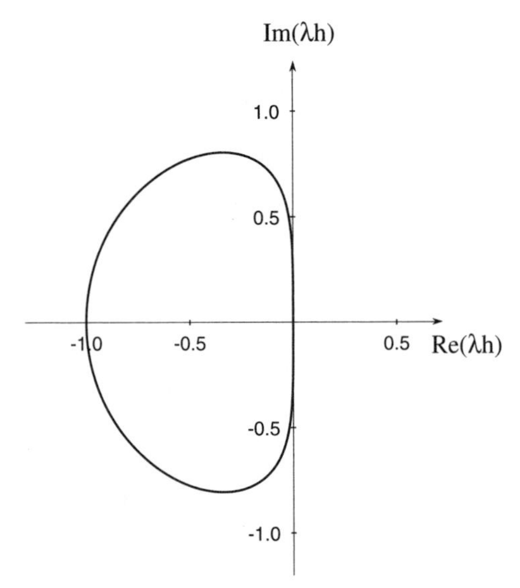

Figure 4.10 Stability diagram for the second-order Adams–Bashforth method.

but it turns out that the instability is very mild. For example, if we use Adams–Bashforth in the problem discussed in Example 5, we obtain $|\sigma_1|^{100} = 1.04$, which is only slightly worse than the second-order Runge–Kutta.

EXAMPLE 4.7 Multi-Step Methods

We solve the problem of Example 3 with the leapfrog and Adams–Bashforth multi-step methods. The details for the leapfrog advancement are given as

$$y_1^{n+1} = y_1^{n-1} + 2hy_2^n$$

$$y_2^{n+1} = y_2^{n-1} - 2h\omega^2 y_1^n.$$

Implementation of the second-order Adams–Bashforth is similar. These multi-step methods are not self-starting and require a single step method to calculate the first time level. Explicit Euler was chosen for the start-up. Once again, numerical results are plotted in Figure 4.11 for $y_o = 1$, $\omega = 4$, and time step $h = 0.15$.

We see that the leapfrog method is stable and with very little amplitude error. There is a slight amplitude error attributed to the explicit Euler calculation for the first time level. This error is not increased by the leapfrog advancement as predicted by our analysis of the model problem. The phase error for leapfrog is seen to be significant and increasing with time. Adams–Bashforth gives a slowly growing numerical solution, which is expected as it is mildly unstable for all problems with purely imaginary eigenvalues.

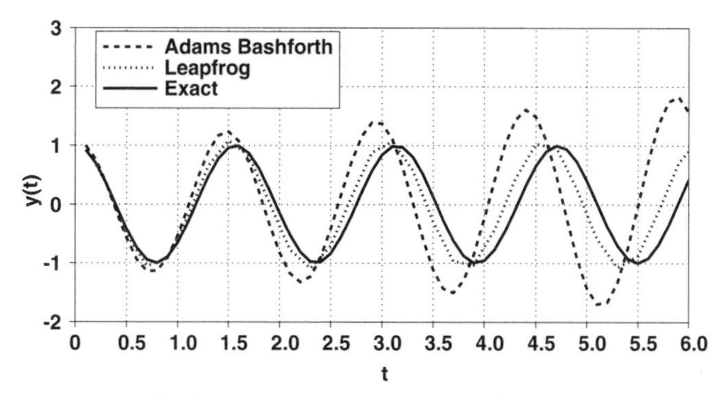

Figure 4.11 Numerical solution of the ODE in Example 3 using multi-step methods.

4.10 System of First-Order Ordinary Differential Equations

Recall that a higher order ordinary differential equations can be converted to a system of first-order ODEs. Systems of ODEs also naturally appear in many physical situations such as chemical reactions among several species or vibration of a complex structure with several elements. A system of ODEs can be written in the generic form

$$\boldsymbol{y}' = \boldsymbol{f}(\boldsymbol{y}, t) \qquad \boldsymbol{y}(0) = \boldsymbol{y}_0 \tag{4.35}$$

where $\boldsymbol{y}$ is a vector with elements y_i and $\boldsymbol{f}(y_1, y_2, y_3, \dots, y_m, t)$ is a vector function with elements $f_i(y_1, y_2, y_3, \dots, y_m, t)$, $i = 1, 2, \dots, m$.

From the applications point of view, numerical solution of a system of ODEs, is a straightforward extension of the techniques used for a single ODE. For example, application of the explicit Euler to (4.35) yields

$$y_i^{(n+1)} = y_i^{(n)} + hf_i\big(y_1^{(n)}, y_2^{(n)}, \dots, y_m^{(n)}, t_n\big) \quad i = 1, 2, 3, \dots, m.$$

The right-hand side can be calculated using data from the previous time step and each equation can be advanced forward.

From the conceptual point of view, there is only one fundamental difference between numerical solution of one ODE and that of a system. This is the *stiffness property* that leads to some numerical problems in systems, but it is not an issue with a single ODE. We shall discuss stiffness in connection with the system of equations with constant coefficients

$$\frac{d\boldsymbol{y}}{dt} = A\boldsymbol{y} \tag{4.36}$$

where A is an $m \times m$ constant matrix. Equation (4.36) is the *model* problem for systems of ODEs. In the same manner that the model equation was helpful in analyzing numerical methods for a single ODE, (4.36) is useful for analyzing numerical methods for systems. From linear algebra we know that this system will have a bounded solution if all the eigenvalues of A have negative real parts.

This is analogous to the single-equation model problem, $y' = \lambda y$, where the real part of λ was negative. Applying the Euler method to (4.36) leads to

$$\mathbf{y}_{n+1} = \mathbf{y}_n + hA\mathbf{y}_n = (I + hA)\mathbf{y}_n$$

or

$$\mathbf{y}_n = (I + hA)^n \mathbf{y}_0.$$

To have a bounded *numerical* solution, the matrix $B^n = (I + hA)^n$ should approach zero for large n. A very important result from linear algebra states:

> The powers of a matrix approach zero for large values of the exponent if the moduli of its eigenvalues are less than 1. That is, if C is a matrix and the moduli of its eigenvalues are less than 1, then
>
> $$\lim_{n\to\infty} C^n \to 0.$$

Therefore, the magnitudes of the eigenvalues of B must be less than 1. The eigenvalues of B are

$$\alpha_i = 1 + h\lambda_i$$

where λ_i are the eigenvalues of the matrix A. Thus, for numerical stability, we must have

$$|1 + \lambda_i h| \leq 1.$$

The eigenvalue with the largest modulus places the most restriction on h. If the eigenvalues are real (and negative), then

$$h \leq \frac{2}{|\lambda|_{\max}}.$$

If the range of the magnitudes of the eigenvalues is large ($|\lambda|_{\max}/|\lambda|_{\min} \gg 1$) and the solution is desired over a large span of the independent variable t, then the system of differential equations is called a *stiff system*. Stiffness arises in physical situations with many degrees of freedom but with widely different rates of responses. Examples include a system composed of two springs, one very stiff and the other very flexible; a mixture of chemical species with very different reaction rates; and a boundary layer (with two disparate length scales).

Stiff systems are associated with numerical difficulties. Problems arise if the system of equations is to be integrated to large values of the independent variable t. Since the step size is limited by the part of the solution with the "fastest" response time (i.e., with the largest eigenvalue magnitude), the number of steps required can become enormous. In other words, even if one is interested only in the long-term behavior of the solution, the time step must still be very small. In practice, to circumvent stiffness, *implicit* methods are used. With implicit methods there is no restriction on the time step due to numerical stability. For high accuracy, one can choose small time steps to resolve the rapidly varying portions of the solution (fast parts) and large time steps in the slowly varying portions. There are stiff ODE solvers (such as *Numerical Recipes*' `stifbs`, MATLAB's `ode23s`, or `lsode`*) that have an adaptive time-step selection mechanism. These are based on implicit methods and automatically reduce or increase the time step depending on the behavior of the solution. Note that with explicit methods one cannot use large time steps in the slowly varying part of the solution. Round-off error will trigger numerical instability associated with the fast part of the solution, even if it is not a significant part of the solution during any portion of the integration period.

EXAMPLE 4.8 A Stiff System (Byrne and Hindmarsh)

The following pair of coupled equations models a ruby laser oscillator

$$\frac{dn}{dt} = -n(\alpha\phi + \beta) + \gamma$$

$$\frac{d\phi}{dt} = \phi(\rho n - \sigma) + \tau(1 + n)$$

with

$$\alpha = 1.5 \times 10^{-18} \qquad \beta = 2.5 \times 10^{-6} \qquad \gamma = 2.1 \times 10^{-6}$$
$$\rho = 0.6 \qquad \sigma = 0.18 \qquad \tau = 0.016$$

and

$$n(0) = -1 \qquad \phi(0) = 0.$$

The variable n represents the population inversion and the variable ϕ represents the photon density. This problem is known to be stiff. We will compare the performance of a stiff equation solution package (`lsode`) with a standard fourth-order Runge–Kutta algorithm. The solution using `lsode` is plotted in Figures 4.12 and 4.13.

Solving the same problem to roughly the same accuracy using a fourth-order Runge–Kutta routine required about 60 times more computer time than the stiff solver. We were unable to use large time steps to improve the efficiency of the Runge–Kutta scheme in the slowly varying portion of the solution because stability is limited by the quickly varying modes in the solution even when they are not very active. The eigenvalue with the highest magnitude still dictates the stability limit even when the modes supported by the smaller eigenvalues are dominating the solution.

* A. C. Hindmarsh, "ODEPACK, a Systematized Collection of ODE Solvers," *Scientific Computing*, edited by R. S. Stepleman et al., (North-Holland, Amsterdam, 1983), p. 55. `lsode` is widely available on the World Wide Web; check for example, `http://www.netlib.org/`.

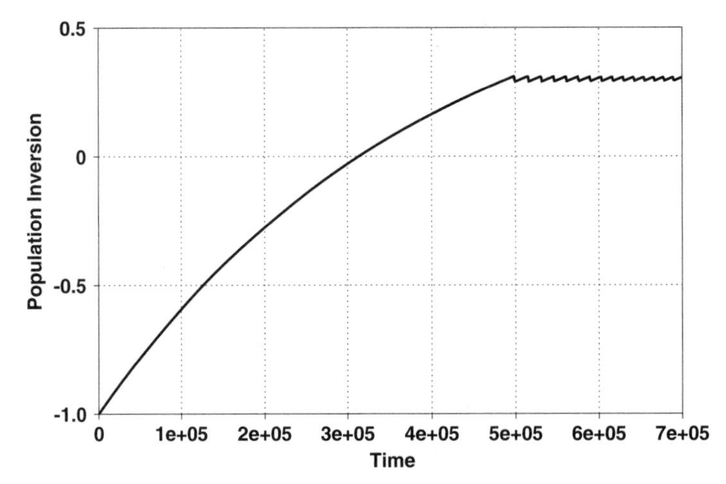

Figure 4.12 Numerical solution of the ODE system in Example 8 using `lsode`.

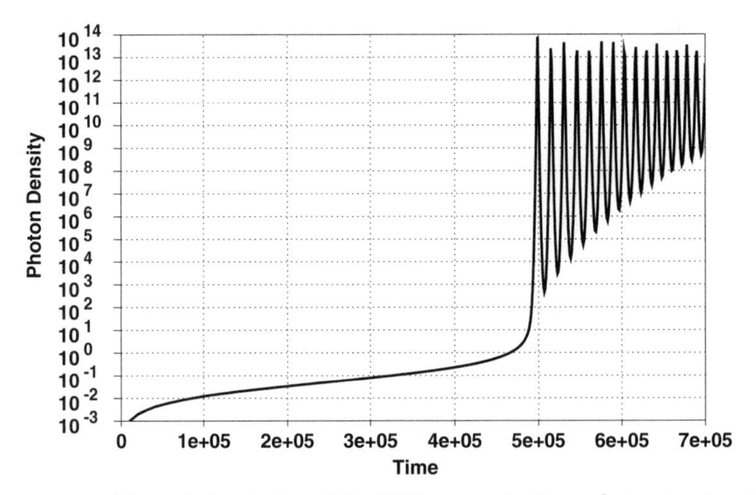

Figure 4.13 Numerical solution of the ODE system in Example 8 using `lsode`.

We have pointed out that the difficulty with implicit methods is that, in general, at each time step, they require solving a non-linear algebraic equation that often requires an iterative solution procedure such as the Newton–Raphson method. It was shown in Section 4.7 that for a single non-linear differential equation, iteration can be avoided by the *linearization technique.* Linearization can also be applied in conjunction with application of implicit methods to a *system* of ODEs. Consider the system

$$\frac{d\boldsymbol{u}}{dt} = \boldsymbol{f}(u_1, u_2, \ldots, u_m, t)$$

where bold letters are used for vectors. Applying the trapezoidal method results in

$$\boldsymbol{u}^{(n+1)} = \boldsymbol{u}^{(n)} + \frac{h}{2}\left[\boldsymbol{f}\left(\boldsymbol{u}^{(n+1)}, t_{n+1}\right) + \boldsymbol{f}\left(\boldsymbol{u}^{(n)}, t_n\right)\right]. \tag{4.37}$$

We would like to linearize $\boldsymbol{f}(\boldsymbol{u}^{(n+1)}, t_{n+1})$. Taylor series expansion of the elements of $\boldsymbol{f}$ denoted by f_i yields

$$f_i\left(\boldsymbol{u}^{(n+1)}, t_{n+1}\right) = f_i\left(\boldsymbol{u}^{(n)}, t_{n+1}\right) + \sum_{j=1}^{m}\left(u_j^{(n+1)} - u_j^{(n)}\right)\frac{\partial f_i}{\partial u_j} + O(h^2)$$

$$i = 1, 2, \ldots, m.$$

We can write this in matrix form as follows:

$$\boldsymbol{f}\left(\boldsymbol{u}^{(n+1)}, t_{n+1}\right) = \boldsymbol{f}\left(\boldsymbol{u}^{(n)}, t_{n+1}\right) + A_n\left(\boldsymbol{u}^{(n+1)} - \boldsymbol{u}^{(n)}\right) + O(h^2)$$

where

$$A_n = \begin{bmatrix} \frac{\partial f_1}{\partial u_1} & \frac{\partial f_1}{\partial u_2} & \cdots & \frac{\partial f_1}{\partial u_m} \\ \vdots & & & \\ \frac{\partial f_m}{\partial u_1} & \frac{\partial f_m}{\partial u_2} & \cdots & \frac{\partial f_m}{\partial u_m} \end{bmatrix}_{(\boldsymbol{u}^{(n)}, t_{n+1})}$$

is the Jacobian matrix. We now substitute this linearization of $\boldsymbol{f}(\boldsymbol{u}^{(n+1)}, t_{n+1})$ into (4.37). It can be seen that, at each time step, instead of solving a non-linear system of algebraic equations, we would solve the following system of *linear* algebraic equations:

$$\left(I - \frac{h}{2}A_n\right)\boldsymbol{u}^{(n+1)} = \left(I - \frac{h}{2}A_n\right)\boldsymbol{u}^{(n)} + \frac{h}{2}\left[\boldsymbol{f}\left(\boldsymbol{u}^{(n)}, t_n\right) + \boldsymbol{f}\left(\boldsymbol{u}^{(n)}, t_{n+1}\right)\right]. \tag{4.38}$$

Note that the matrix A is not constant (its elements are functions of t) and should be updated at every time step.

4.11 Boundary Value Problems

When data associated with a differential equation are prescribed at more than one value of the independent variable, then the problem is a boundary value problem. In initial value problems all the data ($y(0)$, $y'(0)$, ...) are prescribed at one value of the independent variable (in this case at $t = 0$). To have a boundary value problem, we must have at least a second-order differential equation

$$y'' = f(x, y, y') \qquad y(0) = y_0 \qquad y(L) = y_L \tag{4.39}$$

where f is an arbitrary function. Note that here the data are prescribed at $x = 0$ and at $x = L$. The same differential equation, together with data $y(0) = y_0$ and $y'(0) = y_p$, would be an initial value problem.

There are two techniques for solving boundary value problems:

1. *Shooting method.* Shooting is an *iterative technique* which uses the standard methods for initial value problems such as Runge–Kutta methods.

2. *Direct Methods.* These methods are based on straightforward finite-differencing of the derivatives in the differential equation and solving the resulting system of algebraic equations.

We shall begin with the discussion of the shooting method.

4.11.1 Shooting Method

Let's reduce the second-order differential in (4.39) to two first-order equations

$$u = y \qquad v = y'$$
$$\begin{cases} u' = v \\ v' = f(x, u, v). \end{cases} \tag{4.40}$$

The conditions are

$$u(0) = y_0 \quad \text{and} \quad u(L) = y_L.$$

To solve this system (with the familiar methods for initial value problems) one needs one condition for each of the unknowns u and v rather than two for one and none for the other. Therefore, we use a "guess" for $v(0)$ and integrate both equations to $x = L$. At this point, $u(L)$ is compared to y_L; if the agreement is not satisfactory (most likely it will not be unless the user is incredibly lucky), another guess is made for $v(0)$, and the *iterative* process is repeated.

For linear problems this iterative process is very systematic; only two iterations are needed. To illustrate this point, consider the general second-order linear equation

$$y''(x) + A(x)y'(x) + B(x)y(x) = f(x)$$
$$y(0) = y_0 \qquad y(L) = y_L. \tag{4.41}$$

Let's denote two solutions of the equation as $y_1(x)$ and $y_2(x)$, which are obtained using $y_1(0) = y_2(0) = y(0) = y_0$, and two different initial guesses for $y'(0)$. Since the differential equation is linear, the exact solution can be formed as a linear combination of y_1 and y_2

$$y(x) = c_1 y_1(x) + c_2 y_2(x) \tag{4.42}$$

provided that

$$c_1 + c_2 = 1. \tag{4.43a}$$

Next, we require that $y(L) = y_L$, which, in turn, requires that

$$c_1 y_1(L) + c_2 y_2(L) = y_L. \tag{4.43b}$$

Note that $y_1(L)$ and $y_2(L)$ have known numerical values from the solutions $y_1(x)$ and $y_2(x)$, which have already been computed. Equations (4.43) are two

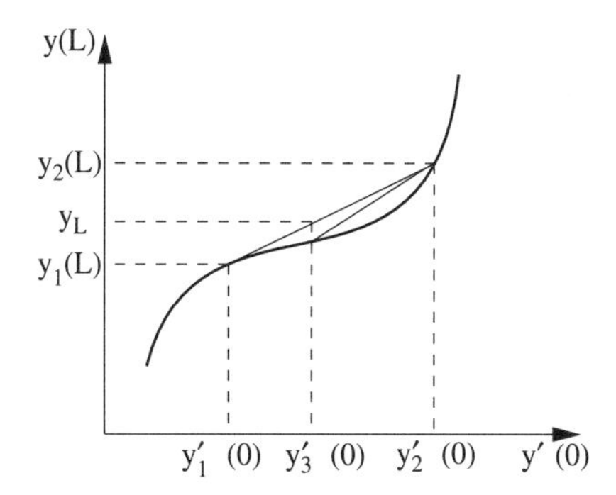

Figure 4.14 Schematic of the functional relationship between $y(L)$ and $y'(0)$. $y_1{}'(0)$ and $y_2{}'(0)$ are the initial guesses leading to $y_1(L)$ and $y_2(L)$ respectively.

linear equations for c_1 and c_2; the solution is

$$c_1 = \frac{y_L - y_2(L)}{y_1(L) - y_2(L)} \quad \text{and} \quad c_2 = \frac{y_1(L) - y_L}{y_1(L) - y_2(L)}.$$

Substitution for c_1 and c_2 into (4.42) gives the desired solution for (4.41). Unfortunately, when (4.39) is non-linear, we may have to perform several iterations to obtain the solution at L to within a prescribed accuracy. Here, we shall demonstrate the solution procedure using the *secant* method which is a well-known technique for the solution of non-linear equations. Consider $y(L)$ as a (non-linear) function of $y'(0)$. This function can be described numerically (and graphically) by several initial guesses for y' and obtaining the corresponding y_L's. A schematic of such a function is shown in Figure 4.14. Suppose that we use two initial guesses, $y_1'(0)$ and $y_2'(0)$, and obtain the solutions $y_1(x)$ and $y_2(x)$ with the values at L denoted by $y_1(L)$ and $y_2(L)$. With the secant method we form the straight line between the points $(y_1'(0), y_1(L))$ and $(y_2'(0), y_2(L))$. This straight line is a crude approximation to the actual curve of $y(L)$ vs. $y'(0)$ between $y_1'(0) \leq y'(0) \leq y_2'(0)$. The equation for this line is

$$y'(0) = y_2'(0) + m[y(L) - y_2(L)],$$

where

$$m = \frac{y_1'(0) - y_2'(0)}{y_1(L) - y_2(L)}$$

is the reciprocal of the slope of the line. The next guess is the value for $y'(0)$ at which the above straight-line approximation to the function predicts y_L. That point is the intersection of the horizontal line from y_L with the straight line, which yields

$$y_3'(0) = y_2'(0) + m[y_L - y_2(L)].$$

In general, the successive iterates are obtained from the formula

$$y'_{\alpha+1}(0) = y'_\alpha(0) + m_{\alpha-1}[y_L - y_\alpha(L)], \tag{4.44a}$$

where $\alpha = 1, 2, 3, \ldots$ is the iteration index and

$$m_{\alpha-1} = \frac{y'_\alpha(0) - y'_{\alpha-1}(0)}{y_\alpha(L) - y_{\alpha-1}(L)} \tag{4.44b}$$

are the reciprocals of the slopes of the successive straight lines (secants). Iterations are continued until $y(L)$ is sufficiently close to y_L. One may encounter difficulty in obtaining a converged solution if $y(L)$ is a very sensitive function of $y'(0)$.

EXAMPLE 4.9 Shooting to Solve the Blasius Boundary Layer

A laminar boundary layer on a flat plate is self-similar and is governed by

$$f''' + ff'' = 0$$

where $f = f(\eta)$ and η is the similarity variable. f and its derivatives are proportional to certain fluid mechanical quantities: $f \propto \Psi$, the stream function; $f' = u/U$, where u is the local fluid velocity and U is the free stream fluid velocity; and $f'' \propto \tau$, the shear stress. Boundary conditions for the equations are derived from the physical boundary conditions on the fluid: "no-slip" at the wall and free stream conditions at large distances from the wall. They are summarized as

$$f'(0) = f(0) = 0 \qquad f'(\infty) = 1.$$

We wish to solve for f and its derivatives throughout the boundary layer. Since one of the boundary conditions is prescribed at $\eta = \infty$ we are required to solve a non-linear *boundary value* problem. Solution proceeds by breaking the third-order problem into a coupled set of first-order equations. Taking $f_1 = f''$, $f_2 = f'$ and $f_3 = f$ gives the following set of ordinary differential equations for the solution:

$$f_1' = -f_1 f_3$$
$$f_2' = f_1$$
$$f_3' = f_2.$$

The solution will be advanced from a prescribed condition at the wall, $\eta = 0$, to $\eta = \infty$. Solutions have been found to converge very quickly for large η and marching from $\eta = 0$ to $\eta = 10$ has been shown to be sufficient for accurate solution. Two conditions are specified at the wall: $f_2 = 0$ and $f_3 = 0$. We must repeatedly solve the whole system and iterate to find the value of $f_1(0)$ that gives the required condition, $f_2 = 1$ at $\eta = \infty$. Two initial guesses were made for $f_1(0)$: $f_1^{(0)}(0) = 1.0$ and $f_1^{(1)}(0) = 0.5$. From these two initial guesses two values for f_2 at "infinity" were calculated: $f_2^{(0)}(10)$ and $f_2^{(1)}(10)$. Starting from these two calculations the secant method may be used to iterate toward an arbitrarily accurate value for $f_1(0)$ based on the following adaptation

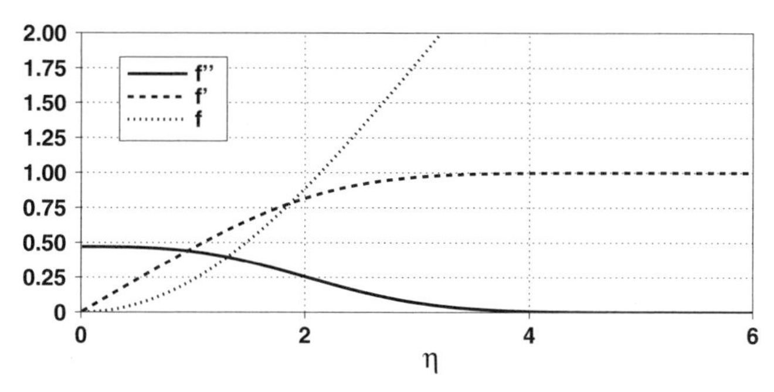

Figure 4.15 Numerical solution of the Blasius boundary layer equation in Example 9.

of (4.44):

$$f_1^{(\alpha+1)}(0) = f_1^{(\alpha)}(0) + \frac{f_1^{(\alpha)}(0) - f_1^{(\alpha-1)}(0)}{f_2^{(\alpha)}(10) - f_2^{(\alpha-1)}(10)}(1 - f_2^{(\alpha)}(10)).$$

Fourth-order Runge–Kutta was used to march the solution from the wall to $\eta = 10$ with a step of $\Delta\eta = 0.01$. Eight secant iterations were necessary after the initial guesses to guarantee convergence to 10 digits. The solutions for f, f', and f'' are plotted in Figure 4.15. We see a "boundary layer shape" in the plot for f' which is the flow velocity. The final solution for $f''(0)$ is $0.469600\ldots$, which agrees with the "accepted" solution.

4.11.2 Direct Methods

With direct methods, one simply approximates the derivatives in the differential equation with a finite difference approximation. The result is a system of algebraic equations for the dependent variables at the node points. For linear differential equations, the system is a linear system of algebraic equations; for non-linear equations, it is a non-linear system of algebraic equations. For example, a second-order approximation to the linear differential equation (4.41) yields

$$\frac{y_{j+1} - 2y_j + y_{j-1}}{h^2} + A_j \frac{y_{j+1} - y_{j-1}}{2h} + B_j y_j = f_j$$

$$y_{(j=0)} = y_0 \qquad y_{(j=N)} = y_L$$

where a uniform grid, $x_j = x_{j-1} + h$, $j = 1, 2, \ldots, N-1$, is introduced between the boundary points x_0 and x_N. Rearranging the terms yields

$$\alpha_j y_{j+1} + \beta_j y_j + \gamma_j y_{j-1} = f_j, \tag{4.45}$$

where

$$\alpha_j = \left(\frac{1}{h^2} + \frac{A_j}{2h}\right) \qquad \beta_j = \left(B_j - \frac{2}{h^2}\right) \qquad \gamma_j = \left(\frac{1}{h^2} - \frac{A_j}{2h}\right)$$

$$j = 1, 2, \ldots, N-1.$$

This is a tridiagonal system of linear algebraic equations. The only special treatment comes at the points next to the boundaries $j = 1$ and $j = N - 1$. At $j = 1$, we have

$$\alpha_1 y_2 + \beta_1 y_1 = f_1 - \gamma_1 y_0.$$

Note that y_0, which is known, is moved to the right-hand side. Similarly, y_N appears on the right-hand side. Thus, the unknowns $y_1, y_2, \ldots, y_{N-1}$ are obtained from the solution of

$$\begin{bmatrix} \beta_1 & \alpha_1 & & & \\ \gamma_2 & \beta_2 & \alpha_2 & & \\ & \ddots & \ddots & \ddots & \\ & & & \gamma_{N-1} & \beta_{N-1} \end{bmatrix} \begin{bmatrix} y_1 \\ y_2 \\ \vdots \\ y_{N-1} \end{bmatrix} = \begin{bmatrix} f_1 - \gamma_1 y_0 \\ f_2 \\ \vdots \\ f_{N-1} - \alpha_{N-1} y_N \end{bmatrix}.$$

Implementation of mixed boundary conditions such as

$$ay(0) + by'(0) = g$$

is also straightforward. For example, one can simply approximate $y'(0)$ with a finite difference approximation such as

$$y'(0) = \frac{-3y_0 + 4y_1 - y_2}{2h} + O(h^2),$$

and solve for y_0 in terms of y_1, y_2, and g. The result is then substituted in the finite difference equation 4.45 evaluated at $j = 1$. Because y_0 now depends on y_1 and y_2, the matrix elements in the first row are also modified. Higher order finite difference approximations can also be used. The only difficulty with higher order methods is that near the boundaries they require data from points outside the domain. The standard procedure is to use lower order approximations for points near the boundary. Moreover, higher order finite differences lead to broader banded matrices instead of a tridiagonal matrix. For example, a pentadiagonal system is obtained with the standard fourth-order central difference approximation to equation (4.41).

Often the solution of a boundary value problem varies rapidly in a part of the domain, and it has a mild variation elsewhere. In such cases it is wasteful to use a fine grid capable of resolving the rapid variations everywhere in the domain. One should use a non-uniform grid spacing (see Section 2.5). In some problems, such as boundary layers in fluid flow problems, the regions of rapid variation are known a priori, and grid points can be clustered where needed. There are also (adaptive) techniques that estimate the grid requirements as the solution progresses and place additional grid points in the regions of rapid variation.

With non-uniform grids one can either use finite difference formulas written explicitly for non-uniform grids or use a coordinate transformation. Both techniques were discussed in Section 2.5. Finite difference formulas for first and second derivatives can be substituted, for example, in (4.41), and the resulting system of equations can be solved. Alternatively, the differential equation can be transformed, and the resulting equation can be solved using uniform mesh formulas.

EXERCISES

1. Consider the equation

$$y' + (2 + 0.01x^2)y = 0$$

$$y(0) = 4 \quad 0 \leq x \leq 10.$$

   (a) Solve this equation using the following numerical schemes: i) Euler, ii) backward Euler, iii) trapezoidal, iv) second-order Runge–Kutta and v) fourth-order Runge–Kutta. Use $\Delta x = 0.1, 0.5, 1.0$ and compare to the exact solution.
   (b) Discuss the stability and accuracy of each scheme.
   (c) For each scheme, estimate the maximum Δx for stable solution (over the given domain) and discuss your estimate in terms of results of part (a).

2. A physical phenomenon is governed by the differential equation

$$\frac{dv}{dt} = -0.2v - 2\cos(2t)v^2$$

   subject to the initial condition $v(0) = 1$.

   (a) Solve this equation analytically.
   (b) Write a program to solve the equation for $0 < t \leq 7$ using the Euler explicit scheme with the following time steps: $h = 0.2, 0.05, 0.025, 0.006$. Plot the four numerical solutions along with the exact solution on one graph. Set the x axis from 0 to 7 and the y axis from 0 to 1.4. Discuss your results.
   (c) In practical problems, the exact solution is not always available. To obtain an accurate solution, we keep reducing the time step (usually by a factor of 2) until two consecutive numerical solutions are nearly the same. Assuming that you do not know the exact solution for the present equation, do you think that the solution corresponding to $h = 0.006$ is accurate (to plotting accuracy)? Justify your answer. In case you find it not accurate enough, obtain a better one.

3. A physical phenomena is governed by a simple differential equation:

$$\frac{dv}{dt} = -\alpha(t)v + \beta(t),$$

   where

$$\alpha(t) = \frac{3t}{(1+t)} \qquad \beta(t) = 2(1+t)^3 e^{-t}.$$

   Assume an initial value $v(0) = 1.0$, and solve the equation for $0 < t < 15$ using the following numerical methods

(a) Euler
(b) Backward Euler
(c) Trapezoidal method
(d) Second-order Runge–Kutta
(e) Fourth-order Runge–Kutta

Try time steps, $h = 0.2,\ 0.8,\ 1.1$. On separate plots, compare your results with the exact solution. Discuss the accuracy and stability of each method. For each scheme, *estimate* the maximum Δt for stable solution (over the given time domain *and* over a very long time).

4. Consider a simple pendulum consisting of mass m attached to a string of length l. The equation of motion for the mass is

$$\theta'' = -\frac{g}{l}\sin\theta,$$

where positive θ is counterclockwise. For small angles θ, $\sin\theta \approx \theta$ and the linearized equation of motion is

$$\theta'' = -\frac{g}{l}\theta.$$

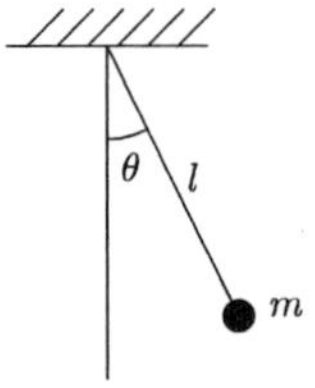

The acceleration due to gravity is $g = 9.81$ m/sec^2, and $l = 0.6$ m. Assume that the pendulum starts from rest with $\theta(t = 0) = 10°$.

(a) Solve the linearized equation for $0 \leq t \leq 6$ using the following numerical methods:
   (i) Euler
   (ii) Backward Euler
   (iii) Second-order Runge-Kutta
   (iv) Fourth-order Runge-Kutta
   (v) Trapezoidal method

   Try time steps, $h = 0.15,\ 0.5,\ 1$. Discuss your results in terms of what you know about the accuracy and stability of these schemes. For each case, and on separate plots, compare your results with the exact solution.

(b) Suppose mass m is placed in a viscous fluid. The linearized equation of motion now becomes

$$\theta'' + c\theta' + \frac{g}{l}\theta = 0.$$

   Let $c = 4$ sec^{-1}. Repeat part (a) with methods (i) and (iii) for this problem. Discuss quantitatively and in detail the stability of your computations as compared to part (a).

(c) Solve the non-linear undamped problem with $\theta(t = 0) = 60°$ with a method of your choice, and compare your results with the corresponding exact linear solution. What steps have you taken to be certain of the accuracy of your results? That is, why should your results be believable? How does the maximum time step for the non-linear problem compare with the prediction of the linear stability analysis?

5. Consider the pendulum problem of Exercise 4. Recall that the linearized equation of motion is

$$\theta'' = -\frac{g}{l}\theta.$$

The pendulum starts from rest with $\theta(t = 0) = 10^\circ$.

(a) Solve the linearized equation for $0 \le t \le 6$ using the following multi-step methods:

(i) Leapfrog
(ii) Second-order Adams–Bashforth

Try time steps, $h = 0.1,\ 0.2,\ 0.5$. Discuss your results in terms of what you know about the accuracy and stability of these schemes. For each case, and on separate plots, compare your results with the exact solution.

(b) The linearized damped equation of motion is

$$\theta'' + c\theta' + \frac{g}{l}\theta = 0.$$

Let $c = 4\ \text{sec}^{-1}$. Repeat part (a) for this problem. Discuss quantitatively and in detail the stability of your computations as compared to part (a). Do your results change significantly using different start-up schemes (e.g., explicit Euler vs. second-order Runge–Kutta)?

6. **Double Pendulum (N. Rott)**

A double pendulum is shown in the figure. One of the pendulums has a space fixed pivot (SFP) and the pivot for the other pendulum (BFP) is attached to the body of the first pendulum. The line connecting the two pivots is of length b and forms an angle β_0 with the vertical, in equilibrium. The total mass of the two elements is m_t, while the BFP pendulum has a mass m_c with a distance c between its center of gravity and its pivot. With m_c concentrated at BFP, the distance of the center of gravity of the total mass from the SFP is a and the moment of inertia of the two bodies is I_t. The moment of inertia of the BFP pendulum about its pivot is I_c. The position angles of the two pendulums with respect to the vertical are α and γ, as shown in the figure.

The equations of motion are (neglecting friction):

$$I_t\ddot{\alpha} + am_t g\ \sin\alpha + bcm_c[C\ddot{\gamma} + S\dot{\gamma}^2] = 0$$

$$I_c\ddot{\gamma} + cm_c g \sin\ \gamma + bcm_c[C\ddot{\alpha} - S\dot{\alpha}^2] = 0$$

where

$$C = \cos\beta_0\ \cos(\alpha - \gamma) - \sin\beta_0\ \sin(\alpha - \gamma)$$

$$S = \sin\beta_0\ \cos(\alpha - \gamma) + \cos\beta_0\ \sin(\alpha - \gamma).$$

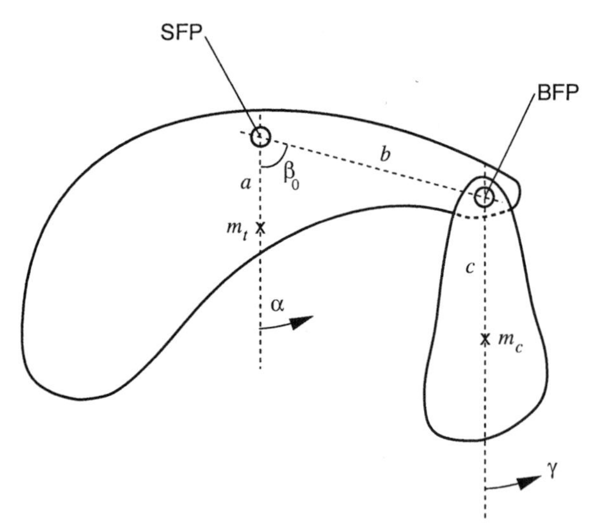

Double pendulum: SFP = space-fixed pivot; BFP = body-fixed pivot.

The following nomenclature is introduced:

$$\frac{am_t g}{I_t} = \lambda^2 \qquad \frac{cm_c g}{I_c} = \omega^2$$

$$\frac{bcm_c}{I_c} = \frac{\omega^2}{g} b = \xi \qquad \frac{bcm_c}{I_t} = \frac{\lambda^2}{g} b \frac{cm_c}{am_t} = \eta.$$

Here λ and ω are the frequencies of the uncoupled modes, while ξ and η are two interaction parameters. Let

$$\beta_0 = \frac{\pi}{2}, \qquad \lambda = 2.74 \text{ rad/s}, \qquad \omega = 5.48 \text{ rad/s},$$

$$\xi = 0.96, \qquad \eta = 0.24.$$

Exchange of Energy

The pendulum system exhibits an interesting coupling when properly "tuned." In a tuned state the modal frequencies are in the ratio 1:2 (here $\omega = 2\lambda$). Then for particular sets of initial conditions, some special interaction takes place in which the two pendulums draw energy from each other at a periodic rate. In that case, when one pendulum oscillates with maximum amplitude, the other stands almost still and the process reverses itself as the energy passes from one pendulum to the other. This phenomenon of energy exchange is periodic if the pendulums are properly tuned. Note that this peculiar motion happens only for well-chosen initial conditions and is usually associated with low energy. Try

$$\alpha_0 = 0, \qquad \dot{\alpha}_0 = 0, \qquad \gamma_0 = \frac{\pi}{12}, \qquad \dot{\gamma}_0 = \pi.$$

Use either your own program, or a canned routine (e.g. *Numerical Recipes*' `odeint` or MATLAB's `ode45`) to solve this system. It is important to experiment with different time steps or tolerance settings (in the canned routines) to ensure that the solution obtained is independent of time step (to plotting accuracy).

Plot the angular deflections (α, γ) and velocities ($\dot{\alpha}, \dot{\gamma}$). Determine the period of energy exchange. Now, pick another set of initial conditions for which periodic energy exchange occurs and find out if the period of energy exchange remains the same. In either case, you should plot the two angles versus time on the same graph in order to reveal the phenomenon of energy exchange. Note that the equations of motion should be solved for a sufficiently long time to exhibit the global periodic nature of the solution.

Chaotic Solution

This system has three degrees of freedom (two angles and two angular velocities make four, but since the system is conservative, the four states are linked in the total energy conservation equation). It is possible for such a system to experience chaotic behavior. Chaotic or unpredictable behavior is usually associated with sensitivity to the initial data. In other words, chaotic behavior implies that two slightly different initial conditions give rise to solutions that differ greatly. In our problem, chaotic solutions are associated with high-energy initial conditions. Try

$$\alpha_0 = \frac{\pi}{2}, \qquad \dot{\alpha}_0 = 5 \text{ rad/s}, \qquad \gamma_0 = 0, \qquad \dot{\gamma}_0 = 0.$$

Simulate the system and plot the two angles versus time. How is the solution different from that of the previous section? Now vary the initial angular velocity $\dot{\alpha}_0$ by 1/2%, i.e. try

$$\alpha_0 = \frac{\pi}{2}, \qquad \dot{\alpha}_0 = 5.025 \text{ rad/s}, \qquad \gamma_0 = 0, \qquad \dot{\gamma}_0 = 0.$$

Plot the angles versus time for the two cases on the same graph and comment on the effect of the small change in the initial conditions. Sensitivity to initial conditions implies sensitivity to truncation and round-off errors as well. Continue your simulations to a sufficiently large time, say $t = 100$ sec, and comment on whether your solution is independent of time step (and hence reliable for large times).

7. Consider the following family of implicit methods for the initial value problem, $y' = f(y)$

$$y_{n+1} = y_n + h[\theta f(y_{n+1}) + (1 - \theta) f(y_n)],$$

where θ is a parameter $0 \leq \theta \leq 1$. The value of $\theta = 1$ yields the backward Euler scheme, and $\theta = 1/2$ yields the trapezoidal method. We have pointed out that not all implicit methods are unconditionally stable. For example, this scheme is conditionally stable for $0 \leq \theta < 1/2$. For the case $\theta = 1/4$, show that the method is conditionally stable, draw its stability diagram, and compare the diagram with the stability diagram of the explicit Euler scheme. Also, plot the stability diagram of the method for $\theta = 3/4$, and discuss possible features of the numerical solution when this method is applied to a problem with a growing exact solution.

8. Non-linear differential equations with several degrees of freedom often exhibit chaotic solutions. Chaos is associated with sensitive dependence to initial conditions; however, numerical solutions are often confined to a so-called strange attractor, which attracts solutions resulting from different initial conditions to its vicinity in the phase space. It is the sensitive dependence on initial conditions that makes many physical systems (such as weather patterns) unpredictable, and it is the attractor that does not allow physical parameters to get out of hand (e.g., very high or low temperatures, etc.) An example of a strange attractor is the Lorenz attractor, which results from the solution of the following equations:

$$\frac{dx}{dt} = \sigma(y - x)$$

$$\frac{dy}{dt} = rx - y - xz$$

$$\frac{dz}{dt} = xy - bz.$$

The values of σ and b are usually fixed ($\sigma = 10$ and $b = 8/3$ in this problem) leaving r as the control parameter. For low values of r, the stable solutions are stationary. When r exceeds 24.74, the trajectories in xyz space become irregular orbits about two particular points.

(a) Solve these equations using $r = 20$. Start from point $(x, y, z) = (1, 1, 1)$, and plot the solution trajectory for $0 \le t \le 25$ in the xy, xz, and yz planes. Plot also x, y, and z versus t. Comment on your plots in terms of the previous discussion.

(b) Observe the change in the solution by repeating (a) for $r = 28$. In this case, plot also the trajectory of the solution in the three-dimensional xyz space (let the z axis be in the horizontal plane; you can use the MATLAB command `plot3(z,y,x)` for this). Compare your plots to (a).

(c) Observe the unpredictability at $r = 28$ by overplotting two solutions versus time starting from two initially nearby points: $(6, 6, 6)$ and $(6, 6.01, 6)$.

9. In this problem we will numerically examine vortex dynamics in two dimensions. We assume that viscosity is negligible, the velocity field is solenoidal ($\nabla \cdot \boldsymbol{u} = 0$), and the vortices may be modeled as potential point vortices. Such a system of potential vortices is governed by a simple set of coupled equations:

$$\frac{dx_j}{dt} = -\frac{1}{2\pi} \sum_{\substack{i=1 \\ i \neq j}}^{N} \frac{\omega_i (y_j - y_i)}{r_{ij}^2} \tag{1a}$$

$$\frac{dy_j}{dt} = \frac{1}{2\pi} \sum_{\substack{i=1 \\ i \neq j}}^{N} \frac{\omega_i (x_j - x_i)}{r_{ij}^2} \tag{1b}$$

where (x_j, y_j) is the position of the jth vortex, ω_j is the strength and rotational direction of the jth vortex (positive ω indicates counter-clockwise rotation),

r_{ij} is the distance between the jth and ith vortices,

$$r_{ij} = \sqrt{(x_i - x_j)^2 + (y_i - y_j)^2}, \tag{2}$$

and N is the number of vortices in the system. For example, in the case of $N = 2$ and $\omega_1 = \omega_2 = 1$, the equations (1a, b) become

$$\frac{dx_1}{dt} = -\frac{1}{2\pi}\frac{(y_1 - y_2)}{r^2} \qquad \frac{dy_1}{dt} = \frac{1}{2\pi}\frac{(x_1 - x_2)}{r^2}$$

$$\frac{dx_2}{dt} = -\frac{1}{2\pi}\frac{(y_2 - y_1)}{r^2} \qquad \frac{dy_2}{dt} = \frac{1}{2\pi}\frac{(x_2 - x_1)}{r^2}$$

$$r = \sqrt{(x_1 - x_2) + (y_1 - y_2)}.$$

Equations (1a) and (1b) may be combined into a more compact form if written for a complex independent variable z_j with $x_j = \text{Real}[z_j]$ and $y_j = \text{Imag}[z_j]$:

$$\frac{dz_j^*}{dt} = \frac{1}{2\pi i}\sum_{\substack{l=1 \\ l\neq j}}^{N} \frac{\omega_l}{z_j - z_l}. \tag{3}$$

The * indicates complex conjugate.

The system has $2N$ degrees of freedom (each vortex has two coordinates that may vary independently). There exist four constraints on the motion of the vortices that may be derived from the flow physics. They are (at a very basic level) conservation of x and y linear momentums, conservation of angular momentum, and conservation of energy. Conservation of energy is useful as it can give a simple measure of the accuracy of a numerical solution. It may be posed as

$$\prod_{j}^{N}\prod_{\substack{i \\ i\neq j}}^{N} \sqrt{r_{ij}} = \text{const.} \tag{4}$$

For $N = 4$ there are four unconstrained degrees of freedom or two unconstrained two-dimensional points of the form (p, q). Such a system may potentially behave chaotically. We will now explore this.

(a) Take $N = 4$ and numerically solve the evolution of the vortex positions. You may solve either Equation (1) or (3). Equation (3) is the more elegant way of doing it but requires a complex ODE solver to be written (same as a real solver but with complex variables). A high-order explicit scheme is recommended (e.g. fourth-order Runge–Kutta). *Numerical Recipes*' `odeint` or MATLAB's `ode45` might be useful. Use as an initial condition $(x, y) = (\pm 1, \pm 1)$; that is, put the vortices on the corners of a square centered at the origin. Take $\omega_j = 1$ for each vortex. Solve for a sufficiently long time to see if the vortex motion is "regular." Use the energy constraint equation (4) to check the accuracy of the solution. Plot the time history of the position of a single vortex in the xy plane.

(b) Perturb one of the initial vortex positions. Move the $(x, y) = (1, 1)$ point to $(x, y) = (1, 1.01)$ and repeat part (a).

(c) Consider a case now where the vortices start on the corners of a rectangle with aspect ratio 2: $(x, y) = (\pm 2, \pm 1)$. Repeat (a).
(d) Again perturb one initial position. Move the $(x, y) = (2, 1)$ point to $(x, y) = (2, 1.01)$ and repeat part (a).
(e) Chaotic systems usually demonstrate a very high dependence upon initial conditions. The solutions from similar but distinct initial conditions often diverge exponentially. Place all vortices in a line: $(x, y)_k = (-1, 0), (\epsilon, 0), (1, 0), (2, 0)$ and accurately solve the problem from time 0 to 4 for $\epsilon = 0$ and $\epsilon = 10^{-4}$. Make a semi-log plot of the distance between the vortices starting at $(0, 0)$ and $(\epsilon, 0)$ versus time for these two runs. Justify the accuracy of the solutions.

10. The following scheme has been proposed for solving $y' = f(y)$:

$$y_{n+1} = y_n + \omega_1 k_1 + \omega_2 k_2,$$

where

$$\begin{aligned} k_1 &= hf(y_n) \\ k_0 &= hf(y_n + \beta_0 k_1) \\ k_2 &= hf(y_n + \beta_1 k_0) \end{aligned}$$

with h being the time step.

(a) Determine the coefficients ω_1, ω_2, β_0, and β_1 that would maximize the order of accuracy of the method. Can you name this method?
(b) Applying this method to $y' = \alpha y$, what is the maximum step size h for α pure imaginary?
(c) Applying this method to $y' = \alpha y$, what is the maximum step size h for α real negative?
(d) With the constraints derived in part (a) draw the stability diagram in the $(h\lambda_R, h\lambda_I)$ plane for this method applied to the model problem $y' = \lambda y$.

11. The following scheme has been proposed for solving $y' = f(y)$:

$$\begin{aligned} y^* &= y^n + \gamma_1 hf(y^n) \\ y^{**} &= y^* + \gamma_2 hf(y^*) + \omega_2 hf(y^n) \\ y^{n+1} &= y^{**} + \gamma_3 hf(y^{**}) + \omega_3 hf(y^*) \end{aligned}$$

where

$$\gamma_1 = 8/15, \qquad \gamma_2 = 5/12, \qquad \gamma_3 = 3/4, \qquad \omega_2 = -17/60,$$
$$\omega_3 = -5/12,$$

with h being the time step.

(a) Give a word description of the method in terms used in this chapter.
(b) What is the order of accuracy of this method?
(c) Applying this method to $y' = \alpha y$, what is the maximum step size h for α pure imaginary and for α negative real?

(d) Draw a stability diagram in the $(h\lambda_R, h\lambda_I)$ plane for this method applied to the model problem $y' = \lambda y$.

12. Chemical reactions often give rise to stiff systems of coupled rate equations. The time history of a reaction of the following form:

$$
\begin{aligned}
&A_1 \rightarrow A_2 \\
&A_2 + A_3 \rightarrow A_1 + A_3 \\
&2A_2 \rightarrow 2A_3
\end{aligned}
$$

is governed by the following rate equations

$$
\begin{aligned}
\dot{C}_1 &= -k_1 C_1 + k_2 C_2 C_3 \\
\dot{C}_2 &= k_1 C_1 - k_2 C_2 C_3 - 2k_3 C_2^2 \\
\dot{C}_3 &= 2k_3 C_2^2
\end{aligned}
$$

where k_1, k_2, and k_3 are reaction rate constants given as

$$k_1 = 0.04, \qquad k_2 = 10.0, \qquad k_3 = 1.5 \times 10^3,$$

and the C_i are the concentrations of species A_i. Initially, $C_1(0) = 0.9$, $C_2(0) = 0.1$, and $C_3(0) = 0$.

(a) What is the analytical steady state solution? Note that these equations should conserve mass, that is, $C_1 + C_2 + C_3 = 1$.

(b) Evaluate the eigenvalues of the Jacobian matrix at $t = 0$. Is the problem stiff?

(c) Solve the given system to a steady state solution ($t = 3000$ represents steady state in this problem) using

  (i) Fourth-order Runge–Kutta (use (b) to estimate the maximum time step).

  (ii) A stiff solver such as *Numerical Recipes*' `stifbs`, `lsode`, or MATLAB's `ode23s`.

Make a log–log plot of the concentrations C_i vs. time. Compare the computer time required for these two methods.

(d) Set up the problem with a linearized trapezoidal method. What advantages would such a scheme have over fourth-order RK?

13. In this problem, we will consider a chemical reaction taking place in our bodies during food digestion. Such chemical reactions are mediated by enzymes, which are biological catalysts. In such a reaction, an enzyme (E) combines with a substrate (S) to form a complex (ES). The ES complex has two possible fates. It can dissociate to E and S or it can proceed to form product P. Such chemical reactions often give rise to stiff systems of coupled rate equations. The time history of this reaction

$$E + S \underset{k_2}{\overset{k_1}{\rightleftharpoons}} ES \overset{k_3}{\longrightarrow} E + P$$

is governed by the following rate equations

$$\frac{dC_S}{dt} = -k_1 C_S C_E + k_2 C_{ES}$$
$$\frac{dC_E}{dt} = -k_1 C_S C_E + (k_2 + k_3) C_{ES}$$
$$\frac{dC_{ES}}{dt} = k_1 C_S C_E - (k_2 + k_3) C_{ES}$$
$$\frac{dC_P}{dt} = k_3 C_{ES}$$

where k_1, k_2, and k_3 are reaction rate constants. The constants for this reaction are

$$k_1 = 2.0 \times 10^3 \qquad k_2 = 1.0 \times 10^{-3} \qquad k_3 = 10.0,$$

and the C_i are the concentrations. Initially, $C_S = 1$, $C_E = 5.0 \times 10^{-5}$, $C_{ES} = 0.0$, $C_P = 0.0$.

(a) Solve the given system of equations to the steady state using:

  (i) Fourth-order Runge–Kutta.

  (ii) A stiff solver such as *Numerical Recipes*' `stifbs`, `lsode`, or MATLAB's `ode23s`.

  Make a log–log plot of the results. Compare the computer time required for these two methods.

(b) Set up and solve the problem with a linearized trapezoidal method. What advantages would such a scheme have over fourth-order RK?

14. Consider the following three-tube model of a kidney (Ivo Babuska)

$$y_1' = a(y_3 - y_1)y_1/y_2$$
$$y_2' = -a(y_3 - y_1)$$
$$y_3' = [b - c(y_3 - y_5) - a y_3(y_3 - y_1)]/y_4$$
$$y_4' = a(y_3 - y_1)$$
$$y_5' = -c(y_5 - y_3)/d$$

where

$$a = 100, \quad b = 0.9, \quad c = 1000, \quad d = 10.$$

Solute and water are exchanged through the walls of the tubes. y_1, y_5, and y_3 represent the concentration of the solute in tubes 1, 2, and 3, respectively. y_2 and y_4 represent the flow rates in tubes 1 and 3. The initial data are

$$y_1(0) = y_2(0) = y_3(0) = 1.0$$
$$y_4(0) = -10, \quad y_5(0) = 0.989$$

(a) Use a stiff ODE solver (such as *Numerical Recipes*' `stifbs`, `lsode`, or MATLAB's `ode23s`) to find the solution for $0 \le t \le 1$. What kind of gradient information did you specify, if any?

(b) Use an explicit method such as the fourth-order Runge–Kutta method and compare the computational effort to that in part (a).
(c) Set up the problem with a second-order implicit scheme with linearization to avoid iterations at each time step.
(d) Solve your setup of part (c). Compare with the other methods. It is advisable to make all your plots on a log–linear scale for this problem.

15. Consider the problem of deflection of a cantilever beam of varying cross section under load P. The differential equation for the deflection y is

$$\frac{d^2}{dx^2}\left(EI\frac{d^2y}{dx^2}\right) = P,$$

where x is the horizontal distance along the beam, E is Young's modulus, and $I(x)$ is the moment of inertia of the cross section. The fixed end of the beam at $x = 0$ implies $y(0) = y'(0) = 0$. At the other end, $x = l$, the bending and shearing moments are zero, that is $y''(l) = y'''(l) = 0$. For the beam under consideration the following data are given:

$$\begin{aligned} I(x) &= 6 \times 10^{-4} e^{-x/l} \text{ m}^4 \\ E &= 230 \times 10^9 \text{ Pa} \\ l &= 5 \text{ m} \\ P &= 10^4 x \text{ N/m}. \end{aligned}$$

Compute the vertical deflection of the beam, $y(x)$. What is the maximum deflection? Where is the maximum curvature in the beam?

It is recommended that you solve this problem using a shooting method. The fourth-order problem should be reduced to a system of four first-order equations in

$$\phi = \begin{bmatrix} y_1 \\ y_2 \\ y_3 \\ y_4 \end{bmatrix} = \begin{bmatrix} y \\ y' \\ y'' \\ y''' \end{bmatrix}.$$

The general solution can be written as

$$\phi = \psi + \sum_{i=1}^{4} c_i \boldsymbol{u}^{(i)}$$

where ψ is the particular solution obtained by shooting with homogeneous conditions. The $\boldsymbol{u}^{(i)}$ are the solutions of the homogeneous equation with initial conditions $\boldsymbol{e}_i$, where the $\boldsymbol{e}_i$ are the Cartesian unit vectors in four dimensions. Show that only three "shots" are necessary to solve the problem and that one only needs to solve a 2×2 system of equations to get c_3 and c_4. In addition, explain why with this procedure only one shot will be necessary for each additional P that may be used.

16. The goal of this problem is to compute the self-similar velocity profile of a compressible viscous flow. The flow is initiated as two adjacent parallel streams that mix as they evolve. After some manipulation and a similarity transformation,

the thin shear layer equations (the boundary layer equations) may be written as the third-order ordinary differential equation:

$$f''' + ff'' = 0 \tag{1}$$

where $f = f(\eta)$, η being the similarity variable. The velocity is given by $f' = u/U_1$, U_1 being the dimensional velocity of the high-speed fluid. U_2 is the dimensional velocity of the low-speed fluid. The boundary conditions are

$$f(0) = 0 \qquad f'(\infty) = 1 \qquad f'(-\infty) = \frac{U_2}{U_1}.$$

This problem is more difficult than the flat-plate boundary layer example in the text because the boundary conditions are specified at three different locations. A very accurate solution, however, may be calculated if you shoot in the following manner:

(a) Guess values for $f'(0)$ and $f''(0)$. These, with the given boundary condition $f(0) = 0$, specify three necessary conditions for advancing the solution numerically from $\eta = 0$. Choose $f'(0) = (U_1 + U_2)/(2U_1)$, the average of the two streams.

(b) Shoot to $\eta = \infty$. (For the purposes of this problem ∞ is 10. This can be shown to be sufficient by asymptotic analysis of the equations.)

(c) Now here's where we get around the fact that we have a three-point boundary value problem. We observe that $g(a\eta) = f(\eta)/a$ also satisfies Equation (1). If we choose $a = f(10)$, which was obtained in (b), the equation recast in g and the corresponding boundary conditions at zero and ∞ are satisfied.

(d) Now take the initial guesses, divide by a and solve for the lower half of the shear layer in the g variable. You have $g(0) = 0$, $g'(0) = f'(0)/a$, and $g''(0) = f''(0)/a$ giving the required initial condition for advancing the solution in g from $\eta = 0$ to $\eta = -10$.

(e) Compare the value of $g'(-10)$ to the boundary condition $f'(-\infty) = U_2/U_1$. Use this difference in a secant method iteration specifying new values of $f''(0)$ until $g'(-10) = U_2/U_1$ is within some error tolerance.

As iteration proceeds, fixing $g'(-10)$ to the boundary condition for $f'(-10)$ in (e) forces a to approach 1 thus making $g \approx f$, the solution. However, a will not actually reach 1, because we do not allow our $f'(0)$ guess to vary. The solution for g, though accurate, may be further refined using step (f).

(f) Use your final value for $g'(0)$ as the fixed $f'(0)$ value in a new iteration. Repeat until you have converged to $a = 1$ and evaluate.

Take $U_1 = 1.0$ and $U_2 = 0.5$, solve, and plot $f'(\eta)$. What was your final value of a? Use an accurate ODE solver for the shooting. (First reproduce the Blasius boundary layer results given in Example 9 in the text. Once that is setup, then try the shear layer.) How different is the solution after (f) than before with $f'(0) = (U_1 + U_2)/(2U_1)$?

17. The diagram shows a body of conical section fabricated from stainless steel immersed in air at a temperature $T_a = 0$. It is of circular cross section that varies

with x. The large end is located at $x = 0$ and is held at temperature $T_A = 5$. The small end is located at $x = L = 2$ and is held at $T_B = 4$.

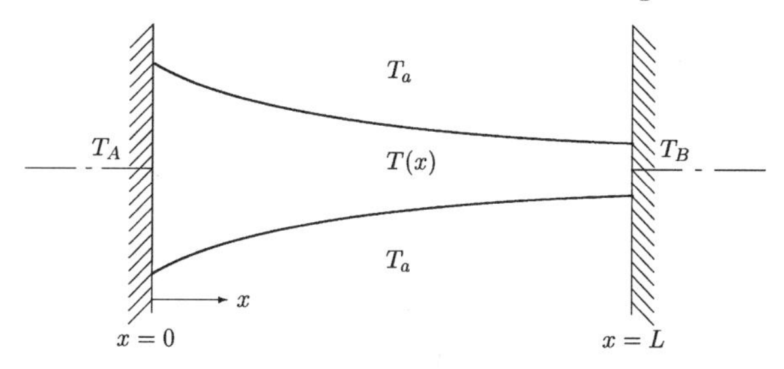

Conservation of energy can be used to develop a heat balance equation at any cross section of the body. When the body is not insulated along its length and the system is at a steady state, its temperature satisfies the following ODE:

$$\frac{d^2T}{dx^2} + a(x)\frac{dT}{dx} + b(x)T = f(x), \tag{1}$$

where $a(x)$, $b(x)$, and $f(x)$ are functions of the cross-sectional area, heat transfer coefficients, and the heat sinks inside the body. In the present example, they are given by

$$a(x) = -\frac{x+3}{x+1}, \qquad b(x) = \frac{x+3}{(x+1)^2}, \qquad \text{and} \quad f(x) = 2(x+1) + 3b(x).$$

(a) In this part, we want to solve (1) using the shooting method.

- (i) Convert the second-order differential equation (1) to a system of 2 first-order differential equations.
- (ii) Use the shooting method to solve the system in (i). Plot the temperature distribution along the body.
- (iii) If the body is insulated at the $x = L$ end, the boundary condition becomes $dT/dx = 0$. In this case use the shooting method to find $T(x)$ and in particular the temperature at $x = L$. Plot the temperature distribution along the body.

(b) We now want to solve (1) directly by approximating the derivatives with finite difference approximations. The interval from $x = 0$ to $x = L$ is discretized using N points (including the boundary points):

$$x_j = \frac{j-1}{N-1}L \quad j = 1, 2, \ldots, N.$$

The temperature at point j is denoted by T_j.

- (i) Discretize the differential equation (1) using the central difference formulas for the second and first derivatives. The discretized equation is valid for $j = 2, 3, \ldots, N-1$ and therefore yields $N-2$ equations for the unknowns $T_1, T_2, \ldots, T_N$.
- (ii) Obtain two additional equations from the boundary conditions ($T_A = 5$ and $T_B = 4$) and write the system of equations in matrix form $AT = f$. Solve this system with $N = 21$. Plot the temperature using symbols on the same plot of part (a)(ii).

FURTHER READING

Dahlquist, G., and Björck, Å. *Numerical Methods*. Prentice-Hall, 1974, Chapter 8.

Forsythe, G. E., Malcolm, M. A., and Moler, C. B. *Computer Methods for Mathematical Computations*. Prentice-Hall, 1977, Chapter 6.

Gear, C. W. *Numerical Initial Value Problems in Ordinary Differential Equations*. Prentice-Hall, 1971.

Press, W. H., Teukolsky, S. A., Vetterling, W. T., and Flannery, B. P. *Numerical Recipes: The Art of Scientific Computing*, Second Edition. Cambridge University Press, 1992, Chapters 16 and 17.

5

Numerical Solution of Partial Differential Equations

Most physical phenomena and processes encountered in engineering problems are governed by partial differential equations, PDEs. Disciplines that use partial differential equations to describe the phenomena of interest include fluid mechanics, where one is interested in predicting the flow of gases and liquids around objects such as cars and airplanes, flow in long distance pipelines, blood flow, ocean currents, atmospheric dynamics, air pollution, underground dispersion of contaminants, plasma reactors for semiconductor equipments, and flow in gas turbine and internal combustion engines. In solid mechanics, problems encountered in vibrations, elasticity, plasticity, fracture mechanics, and structure loadings are governed by partial differential equations. The propagation of acoustic and electromagnetic waves, and problems in heat and mass transfer are also governed by partial differential equations.

Numerical simulation of partial differential equations is far more demanding than that of ordinary differential equations. Also the diversity of types of partial differential equations precludes the availability of general purpose "canned" computer programs for their solutions. Although commercial codes are available in different disciplines, the user must be aware of the workings of these codes and/or perform some complementary computer programming and have a basic understanding of the numerical issues involved. However, with the advent of faster computers, numerical simulation of physical phenomena is becoming more practical and more common. Computational prototyping is becoming a significant part of the design process for engineering systems. With ever increasing computer performance the outlook is even brighter, and computer simulations are expected to replace expensive physical testing of design prototypes.

In this chapter we will develop basic numerical methods for the solution of PDEs. We will consider both initial (transient) and equilibrium problems. We will begin by demonstrating that numerical methods for PDEs are straightforward extensions of methods developed for initial and boundary value problems in ODEs.

5.1 Semi-Discretization

A partial differential equation can be readily converted to a system of ordinary differential equations by using finite difference approximations for derivatives in all but one of the dimensions. Consider, for example, the one-dimensional diffusion equation for $\phi(x, t)$:

$$\frac{\partial \phi}{\partial t} = \alpha \frac{\partial^2 \phi}{\partial x^2}. \tag{5.1}$$

Suppose the boundary and initial conditions are

$$\phi(0, t) = \phi(L, t) = 0 \quad \text{and} \quad \phi(x, 0) = g(x).$$

We discretize the coordinate x with $N + 1$ uniformly spaced grid points

$$x_j = x_{j-1} + \Delta x \quad j = 1, 2, \ldots, N.$$

The boundaries are at $j = 0$ and $j = N$, and $j = 1, 2, \ldots, N - 1$ represent the interior points. If we use the second-order central difference scheme to approximate the second derivative in (5.1) we get

$$\frac{d\phi_j}{dt} = \alpha \frac{\phi_{j+1} - 2\phi_j + \phi_{j-1}}{\Delta x^2} \quad j = 1, 2, 3, \ldots, N - 1 \tag{5.2}$$

where $\phi_j = \phi(x_j, t)$. This is a system of $N - 1$ *ordinary* differential equations that can be written in matrix form as

$$\frac{d\phi}{dt} = A\phi, \tag{5.3}$$

where ϕ_j are the (time-dependent) elements of the vector $\phi(t)$, and A is an $(N - 1) \times (N - 1)$ tridiagonal matrix:

$$A = \frac{\alpha}{\Delta x^2} \begin{bmatrix} -2 & 1 & & & & \\ 1 & -2 & 1 & & & \\ & & \ddots & \ddots & \ddots & \\ & & & 1 & -2 & 1 \\ & & & & 1 & -2 \end{bmatrix}.$$

Since A is a banded matrix, it is sometimes denoted using the compact notation

$$A = \frac{\alpha}{\Delta x^2} B[1, -2, 1].$$

We have now completed semi-discretization of the partial differential equation (5.1). The result is a system of ordinary differential equations that can be solved using any of the numerical methods introduced for ODEs such as Runge–Kutta formulas or multi-step methods. However, when dealing with systems, we have to be concerned about stiffness (Section 4.10). Recall that the range of the eigenvalues of A determines whether the system is stiff. Fortunately for certain banded matrices, analytical expressions are available for the eigenvalues and eigenvectors. For example, eigenvalues of A can be obtained from a known formula for the eigenvalues of a tridiagonal matrix with constant entries. Note that the diagonal and sub-diagonals of A are -2, 1, and 1 respectively, which do not change throughout the matrix. This result is described in the following exercise from linear algebra.

EXERCISE

Let T be an $(N-1) \times (N-1)$ tridiagonal matrix, $B[a, b, c]$. Let $D_{(N-1)}$ be the determinant of T.

(i) Show that $D_{(N-1)} = bD_{(N-2)} - acD_{(N-3)}$.
(ii) Show that $D_{(N-1)} = r^{(N-1)}/\sin\theta \sin(N\theta)$, where $r = \sqrt{ac}$ and $2r\cos\theta = b$. *Hint: Use induction.*
(iii) Show that the eigenvalues of A are given by

$$\lambda_j = b + 2\sqrt{ac}\cos\alpha_j, \tag{5.4}$$

where

$$\alpha_j = \frac{j\pi}{N} \quad j = 1, 2, \ldots, N-1.$$

Therefore, according to this result, the eigenvalues of A are

$$\lambda_j = \frac{\alpha}{\Delta x^2}\left(-2 + 2\cos\frac{\pi j}{N}\right) \quad j = 1, 2, \ldots, N-1.$$

The eigenvalue with the smallest magnitude is

$$\lambda_1 = \frac{\alpha}{\Delta x^2}\left(-2 + 2\cos\frac{\pi}{N}\right).$$

For large N, the series expansion for $\cos(\pi/N)$,

$$\cos\frac{\pi}{N} = 1 - \frac{1}{2!}\left(\frac{\pi}{N}\right)^2 + \frac{1}{4!}\left(\frac{\pi}{N}\right)^4 + \cdots,$$

converges rapidly. Retaining the first two terms in the expansion results in

$$\lambda_1 \approx -\frac{\pi^2\alpha}{N^2\Delta x^2}. \tag{5.5}$$

Also, for large N we have

$$\lambda_{N-1} \approx -4\frac{\alpha}{\Delta x^2}. \tag{5.6}$$

Therefore, the ratio of the eigenvalue with the largest modulus to the eigenvalue with the smallest modulus is

$$\left|\frac{\lambda_{N-1}}{\lambda_1}\right| \approx \frac{4N^2}{\pi^2}.$$

Clearly, for large N the system is stiff.

The knowledge of the eigenvalues also provides insight into the physical behavior of the numerical solution. Notice that all the eigenvalues of A are real and negative. To see how the eigenvalues enter into the solution of (5.3), we diagonalize A using the standard eigenvector diagonalization procedure from linear algebra (Appendix); i.e., let

$$A = S\Lambda S^{-1}, \tag{5.7}$$

where $\Lambda = S^{-1}AS$ is the diagonal matrix with the eigenvalues of A on the diagonal; S is the matrix whose columns are the eigenvectors of A. Note that since A is symmetric, we are always guaranteed to have a set of orthogonal eigenvectors, and the decomposition in (5.7) is always possible. Substituting this decomposition for A into (5.3) yields

$$\frac{d\psi}{dt} = \Lambda\psi, \tag{5.8}$$

where $\psi = S^{-1}\psi$. Since Λ is diagonal the equations are uncoupled and the solution can be obtained readily

$$\psi_j(t) = \psi_j(0)e^{\lambda_j t} \tag{5.9}$$

where $\psi_j(0)$ can be obtained in terms of the original initial conditions from $\psi(0) = S^{-1}\phi(0)$. The solution for the original variable is $\phi = S\psi$, which can be written as (see Appendix)

$$\phi = \psi_1 S^{(1)} + \psi_2 S^{(2)} + \cdots + \psi_{N-1} S^{(N-1)}, \tag{5.10}$$

where $S^{(j)}$ is the jth column of the matrix of eigenvectors S. Note that the solution consists of a superposition of several "modes"; the eigenvalues of A determine the temporal behavior of the solution (according to (5.9)) and its eigenvectors determine its spatial behavior. A key result of this analysis is that *the negative real eigenvalues of A result in a decaying solution in time*, which is the expected behavior for the diffusion equation. The rate of decay is related to the magnitude of the eigenvalues.

EXAMPLE 5.1 Heat Equation

We will examine the stability of numerical solutions of the inhomogeneous heat equation

$$\frac{\partial T}{\partial t} = \alpha\frac{\partial^2 T}{\partial x^2} + (\pi^2 - 1)e^{-t}\sin\pi x \qquad 0 \le x \le 1; t \ge 0,$$

with the initial and boundary conditions

$$T(0,t) = T(1,t) = 0 \qquad T(x,0) = \sin\pi x.$$

As shown in this section, this equation is first discretized in space using the second-order central difference scheme resulting in the following coupled set of ordinary differential equations with time as the independent

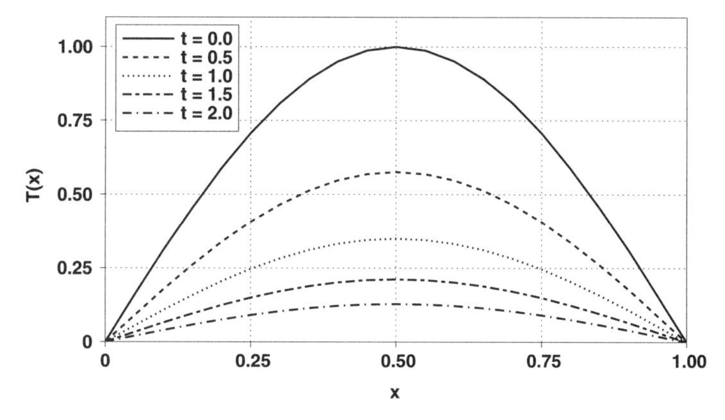

Figure 5.1 Numerical solution of the heat equation in Example 5.1 using $\Delta t = 0.001$.

variable:

$$\frac{d\boldsymbol{T}}{dt} = \frac{\alpha}{\Delta x^2} B[1, -2, 1]\boldsymbol{T} + \boldsymbol{f}.$$

The vector $\boldsymbol{f}$ is the inhomogeneous term and has the components

$$f_j = (\pi^2 - 1)e^{-t} \sin \pi x_j.$$

Note that if non-zero boundary conditions were prescribed, then the known boundary terms would move to the right-hand side, resulting in a change in f_1 and f_{N-1}. Recall that the PDE has been converted to a set of ODEs. Therefore, the stability of the numerical solution depends upon the eigenvalue of the system having the largest magnitude, which is known (from (5.6)) to be

$$\lambda_{N-1} \approx -4\frac{\alpha}{\Delta x^2}.$$

Suppose we wish to solve this equation with the explicit Euler scheme. We know from Section 4.10 that for real and negative λ

$$\Delta t_{\max} = \frac{2}{|\lambda|_{\max}} = \frac{\Delta x^2}{2\alpha}.$$

Taking $\alpha = 1$ and $\Delta x = 0.05$ (giving $N = 21$ grid point over the x domain), we calculate $\Delta t_{\max} = 0.00125$. Results for $\Delta t = 0.001$ are plotted in Figure 5.1.

The numerical solution is decaying as predicted. On the other hand, selecting $\Delta t = 0.0015$ gives the numerical solution shown in Figure 5.2, which is clearly unstable as predicted by the stability analysis.

Now, let us consider a semi-discretization of the following first-order wave equation

$$\frac{\partial u}{\partial t} + c\frac{\partial u}{\partial x} = 0 \qquad 0 \le x \le L \qquad t \ge 0, \tag{5.11}$$

with the boundary condition $u(0, t) = 0$. This is a simple model equation for

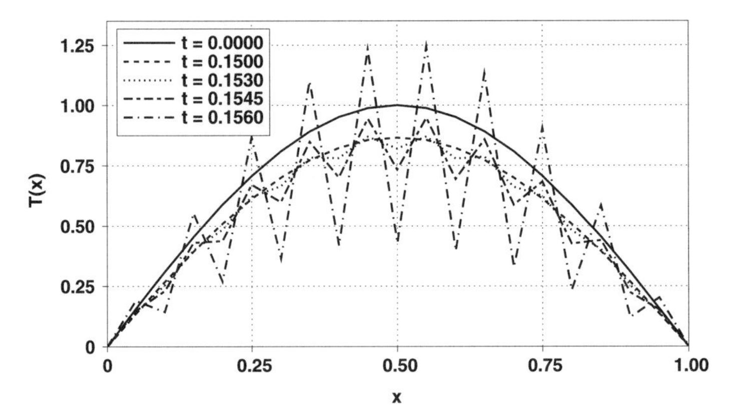

Figure 5.2 Numerical solution of the heat equation in Example 5.1 using $\Delta t = 0.0015$.

the convection phenomenon. The exact solution of this equation is such that an initial disturbance in the domain (as prescribed in the initial condition $u(x, 0)$) simply propagates with the constant convection speed c in the positive or negative x direction depending on the sign of c. For the present case, we assume that $c > 0$. Semi-discretization with the central difference formula leads to

$$\frac{du_j}{dt} + c\frac{u_{j+1} - u_{j-1}}{2\Delta x} = 0. \tag{5.12}$$

In matrix notation we have

$$\frac{d\boldsymbol{u}}{dt} = -\frac{c}{2\Delta x}B\boldsymbol{u}$$

where, $B = B[-1, 0, 1]$ is a tridiagonal matrix with 0's on the diagonal and -1's and 1's for the sub- and super-diagonals respectively. From analytical considerations, no boundary condition is prescribed at $x = L$, however, a special numerical boundary treatment is required at $x = L$ owing to the use of central spatial differencing in the problem. A typical well behaved *numerical* boundary treatment at $x = L$ slightly modifies the last row of B, but for the present discussion we are not going to concern ourselves with this issue. Using (5.4), the eigenvalues of B are

$$\lambda_j = -\frac{c}{\Delta x}\left(i \cos\frac{\pi j}{N}\right) \quad j = 1, 2, \ldots, N-1$$

where, we have assumed that B is $(N-1) \times (N-1)$. Thus, the eigenvalues of the matrix resulting from semi-discretization of the *convection equation*, (5.11), are purely imaginary, i.e., $\lambda_j = i\omega_j$, where, $\omega_j = -\frac{c}{\Delta x}(\cos\frac{\pi j}{N})$. An eigenvector decomposition analysis similar to that done above for the diffusion equation leads to the key conclusion that the solution is a superposition of modes, where each mode's temporal behavior is given by $e^{i\omega_j t}$, which has oscillatory (non-decaying) or sinusoidal character.

This is a good place to pause and reflect on the important results deduced from semi-discretization of two important equations. Spatial discretizations of (5.1) and (5.11) have led to important insights into the behaviors of the respective solutions. These two equations are examples of two limiting cases, one with a decaying solution (*negative real eigenvalues*) and the other with oscillatory behavior (*imaginary eigenvalues*). Diagonalizations of the matrices arising from discretizations uncoupled the systems into equations of the form

$$y' = \lambda y.$$

This of course, is the familiar model equation used in Chapter 4 for the analysis of numerical methods for ordinary differential equations. This model acts as an important bridge between numerical methods for ODEs and the time advancement schemes for PDEs. It is through this bridge that virtually all the results obtained for ODEs will be directly applicable to the numerical solution of time-dependent PDEs.

Recall that the analysis of ODEs was performed for complex λ. In the case of ODEs we argued that λ must be complex to model sinusoidal behavior arising from higher order ODEs. Here we see that the real and imaginary parts of λ model two very different physical systems, namely diffusion and convection. The case with λ real and negative is a model for the partial differential equation (5.1), and the case with λ purely imaginary is a model for (5.11). Thus, when applying standard time-step marching methods to these partial differential equations, the results derived for ODEs should be applicable. For example, recall that application of the Euler scheme to $y' = \lambda y$ was unstable for purely imaginary λ. Thus, we can readily deduce that *application of the explicit Euler to the convection equation (5.11), with second-order central spatial differencing (5.12), will lead to an unconditionally unstable numerical solution*, and the application of the same scheme to the heat equation (5.1) is conditionally stable.

In the heat equation case, the maximum time step is obtained from the requirement (Section 4.10)

$$|1 + \Delta t \lambda_i| \leq 1 \quad i = 1, 2, 3, \ldots, N-1,$$

which leads to

$$\Delta t \leq \frac{2}{|\lambda|_{\max}}$$

where $|\lambda|_{\max}$ is the magnitude of the eigenvalue with the largest modulus of the matrix obtained from semi-discretization of (5.1). Using the expression for this largest eigenvalue given in (5.6) leads to

$$\Delta t \leq \frac{\Delta x^2}{2\alpha}. \tag{5.13}$$

This is a rather severe restriction on the time step. It implies that increasing the spatial accuracy (reducing Δx) must be accompanied by a significant reduction in the time step.

EXAMPLE 5.2 Convection Equation

We consider numerical solutions of the homogeneous convection equation

$$\frac{\partial u}{\partial t} + c\frac{\partial u}{\partial x} = 0 \quad x \geq 0,\ t \geq 0,$$

with the initial and boundary conditions

$$u(0,t) = 0 \qquad u(x,0) = e^{-200(x-0.25)^2}.$$

Although the proper spatial domain for this partial differential equation is semi-infinite as indicated earlier, numerical implementation requires a finite domain. Thus, for this example, we arbitrarily truncate the domain to $0 \leq x \leq 1$. Numerical formulation starts by first discretizing the PDE in space using a second-order central difference scheme, giving the following system of coupled ordinary differential equations

$$\frac{d\boldsymbol{u}}{dt} = -\frac{c}{\Delta x}B[-1,0,1]\boldsymbol{u}.$$

The coefficient matrix on the right hand side is a skew-symmetric matrix and therefore has purely imaginary eigenvalues. Explicit Euler is unstable for systems with purely imaginary eigenvalues, and therefore we expect an unconditionally unstable solution if explicit Euler is used for the time marching scheme in this problem. Nevertheless, we will attempt a numerical solution using second-order central differencing in the interior of the domain. A one-sided differencing scheme is used on the right boundary to allow the waves to pass smoothly out of the computational domain. The solution with $c = 1$, $\Delta x = 0.01$, and $\Delta t = 0.01$ is plotted in Figure 5.3. We see that the numerical solution is indeed unstable and the instability

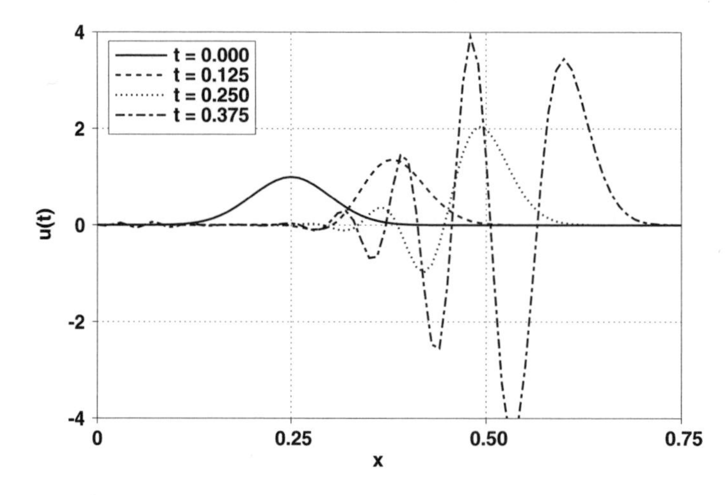

Figure 5.3 Numerical solutions of the convection equation in Example 5.2 using the explicit Euler time advancement and second-order central difference in space.

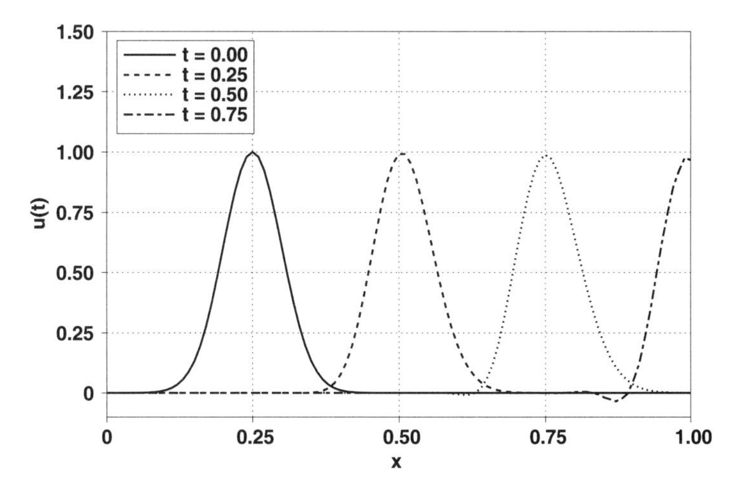

Figure 5.4 Numerical solutions of the convection equation in Example 5.2 using fourth-order Runge–Kutta time advancement, and second-order central difference in space.

sets in even before the disturbance reaches the artificial outflow boundary at $x = 1$.

The stability diagram for the fourth-order Runge–Kutta scheme includes a portion of the imaginary axis (see Figure 4.8) and therefore, we expect this method to be conditionally stable for the convection equation considered in this example (having purely imaginary eigenvalues). Results of a fourth-order Runge–Kutta calculation with $c = 1$, $\Delta x = 0.01$, and $\Delta t = 0.01$ are given in Figure 5.4.

This appears to be an accurate solution, showing the initial disturbance propagating out of the computational domain with only a small amplitude error which could be reduced by refining the time step and/or the spatial grid spacing. We will further discuss our choice of the time step for this example in the following sections.

5.2 von Neumann Stability Analysis

The preceding stability analysis uses the eigenvalues of the matrix obtained from a semi-discretization of the partial differential equation at hand. Different spatial differencing schemes lead to different stability criteria for a given time advancement scheme. We shall refer to this type of analysis as the *matrix stability analysis*. Since boundary conditions are implemented in the semi-discretization, their effects are accounted for in the matrix stability analysis. The price paid for this generality is the need to know the eigenvalues of the matrix that arises from the spatial discretization. Unfortunately, analytical expressions for the eigenvalues are only available for very simple matrices, and therefore, the matrix stability analysis is not widely used.

Experience has shown that in most cases, numerical stability problems arise solely from the (full) discretization of the partial differential equation inside the domain and not from the boundary conditions. von Neumann's stability analysis is a widely used (back of an envelope) analytical procedure for determining the stability properties of a numerical method applied to a PDE that does not account for the effects of boundary conditions. In fact, it is assumed that the boundary conditions are periodic; that is, the solution and its derivatives are the same at the two ends of the domain. The technique works for linear, *constant coefficient* differential equations that are discretized on *uniformly spaced* spatial grids.

Let's demonstrate von Neumann's technique by applying it to the discrete equation

$$\phi_j^{(n+1)} = \phi_j^{(n)} + \frac{\alpha \Delta t}{\Delta x^2}\left(\phi_{j+1}^{(n)} - 2\phi_j^{(n)} + \phi_{j-1}^{(n)}\right). \tag{5.14}$$

Equation (5.14) results from approximating the spatial derivative in (5.1) with the second-order central difference and using the explicit Euler for time advancement. The key part of von Neumann's analysis is to assume a solution of the form

$$\phi_j^{(n)} = \sigma^n e^{ikx_j} \tag{5.15}$$

for the discrete equation (5.14). Note that the assumption of spatial periodicity is already worked into the form of the solution in (5.15); the period is $2\pi/k$. To check whether this solution works, we substitute (5.15) into (5.14) and obtain

$$\sigma^{n+1} e^{ikx_j} = \sigma^n e^{ikx_j} + \frac{\alpha \Delta t}{\Delta x^2}\sigma^n\left(e^{ikx_{j+1}} - 2e^{ikx_j} + e^{ikx_{j-1}}\right).$$

Noting that

$$x_{j+1} = x_j + \Delta x \quad \text{and} \quad x_{j-1} = x_j - \Delta x$$

and dividing both sides by $\sigma^n e^{ikx_j}$ leads to

$$\sigma = 1 + \left(\frac{\alpha \Delta t}{\Delta x^2}\right)[2\cos(k\Delta x) - 2]. \tag{5.16}$$

For stability, we must have $|\sigma| \leq 1$ (otherwise, σ^n in (5.15) would grow unbounded):

$$\left|1 + \left(\frac{\alpha \Delta t}{\Delta x^2}\right)[2\cos(k\Delta x) - 2]\right| \leq 1.$$

In other words, we must have

$$-1 \leq 1 + \left(\frac{\alpha \Delta t}{\Delta x^2}\right)[2\cos(k\Delta x) - 2] \leq 1.$$

The right-hand inequality is always satisfied since $[2\cos(k\Delta x) - 2]$ is always less than or equal to zero. The left-hand inequality can be recast as

$$\left(\frac{\alpha\Delta t}{\Delta x^2}\right)[2\cos(k\Delta x) - 2] \geq -2$$

or

$$\Delta t \leq \frac{\Delta x^2}{\alpha[1 - \cos(k\Delta x)]}.$$

The worst (or the most restrictive) case occurs when $\cos(k\Delta x) = -1$. Thus, the time step is limited by

$$\Delta t \leq \frac{\Delta x^2}{2\alpha}.$$

This is identical to (5.13), which was obtained using the matrix stability analysis. However, the agreement is just a coincidence; in general, there is no reason to expect such perfect agreement between the two methods of stability analysis (each of which assumed different boundary conditions for the same PDE).

In summary, the von Neumann analysis is an *analytical* technique that is applied to the full (space–time) discretization of a partial differential equation. The technique works whenever the space-dependent terms are eliminated after substituting the periodic form of the solution given in (5.15). For example, if in (5.1), α were a known function of x, then the von Neumann analysis would not, in general, work. In this case σ would have to be a function of x and the simple solution given in (5.16) would no longer be valid. The same problem would arise if a *non-uniformly* spaced spatial grid were used. Of course, in these cases the matrix stability analysis would still work, but (for variable α or non-uniform meshes) the eigenvalues would not be available via an analytical formula such as (5.4) moreover, one would have to resort to well-known numerical techniques to estimate the eigenvalue with the highest magnitude for a given N. However, in case such an estimate is not available, experience has shown us that using the maximum value of $\alpha(x)$ and/or the smallest Δx in (5.13) gives an adequate estimate for $\Delta t_{\max}$.

5.3 Modified Wavenumber Analysis

In Section 2.3 the accuracies of finite difference operators were evaluated by numerically differentiating e^{ikx} and comparing their modified wavenumbers with the exact wavenumber. In this section, the modified wavenumbers of differencing schemes are used in the analysis of the stability characteristics of numerical solutions of partial differential equations. This is the third method of stability analysis for PDEs discussed in this chapter.

The modified wavenumber analysis is very similar to the von Neumann analysis; in many ways it is more straightforward. It is intended to readily expand the range of applicability of what we have learned about the stability properties of a time-marching scheme for ordinary differential equations to the application of the same time-advancement method to partial differential equations.

Consider the heat equation (5.1). Assuming a solution of the form

$$\phi(x, t) = \psi(t)e^{ikx}$$

and substituting into (5.1) leads to

$$\frac{d\psi}{dt} = -\alpha k^2 \psi. \tag{5.17}$$

In the assumed form of the solution, k is the wavenumber. In practice, instead of using the analytical differentiation that led to (5.17), one uses a finite difference scheme to approximate the spatial derivative. For example, using the second-order central finite difference scheme, we have

$$\frac{d\phi_j}{dt} = \alpha \frac{\phi_{j+1} - 2\phi_j + \phi_{j-1}}{\Delta x^2} \qquad j = 1, 2, 3, \ldots, N - 1. \tag{5.18}$$

Let's assume that

$$\phi_j = \psi(t)e^{ikx_j}$$

is the solution for the (semi-) discretized equation (5.18). Substitution in (5.18) and division by e^{ikx_j} leads to

$$\frac{d\psi}{dt} = -\frac{2\alpha}{\Delta x^2}[1 - \cos(k\Delta x)]\psi$$

or

$$\frac{d\psi}{dt} = -\alpha k'^2 \psi, \tag{5.19}$$

where

$$k'^2 = \frac{2}{\Delta x^2}[1 - \cos(k\Delta x)].$$

By analogy to equation (5.17), k' is called the *modified* wavenumber, which was first introduced in Section 2.3. Application of any other finite difference scheme instead of the second-order scheme used here would have also led to the same form as (5.19), but with a different modified wavenumber. As discussed in Section 2.3, each finite difference scheme has a distinct modified wavenumber associated with it.

Now, we can apply our knowledge of numerical analysis of ODEs to (5.19). The key observation is that (5.19) is identical to the model ordinary differential equation $y' = \lambda y$, with $\lambda = -\alpha k'^2$. In Chapter 4, we extensively studied the stability properties of various numerical methods for ODEs with respect to this model equation. Now, using the modified wavenumber analysis, we can readily obtain the stability properties of any of those time-advancement methods when applied to a partial differential equation. All we have to do is replace λ with $-\alpha k'^2$ in our ODE analysis. For example, recall from Section 4.3 that when the explicit Euler method was applied to $y' = \lambda y$, with λ real and negative, the time step was bounded by

$$\Delta t \leq \frac{2}{|\lambda|}.$$

For the heat equation, this result is used as follows. If the explicit Euler time-marching scheme is applied to the partial differential equation (5.1) in conjunction with the second-order central difference for the spatial derivative, the time step should be bounded by

$$\Delta t \leq \frac{2}{\frac{2\alpha}{\Delta x^2}[1 - \cos(k\Delta x)]}.$$

The *worst case scenario* (i.e., the maximum limitation on the time step) occurs when $\cos(k\Delta x) = -1$, which leads to (5.13), which was obtained with the von Neumann analysis.

The advantage of the modified wavenumber analysis is that the stability limits for different time-advancement schemes applied to the same equation are readily obtained. For example, if instead of the explicit Euler we had used a fourth-order Runge–Kutta scheme, the stability limit would have been

$$\Delta t \leq \frac{2.79\Delta x^2}{4\alpha},$$

which is obtained directly from the intersection of the stability diagram for the fourth-order Runge–Kutta with the real axis (see Figure 4.8). Similarly, since $-\alpha k'^2$ is real and negative, it is readily deduced that *application of the leapfrog scheme to* (5.1) *would lead to numerical instability*.

As a further illustration of the modified wavenumber analysis, consider the convection equation (5.11). Suppose, the second-order central difference scheme is used to approximate the spatial derivative. In the wavenumber space (which we reach by assuming solution of the form $\phi_j = \psi(t)e^{ikx_j}$), the semi-discretized equation is written as

$$\frac{d\psi}{dt} = -ik'c\psi, \tag{5.20}$$

where

$$k' = \frac{\sin(k\Delta x)}{\Delta x} \tag{5.21}$$

is the modified wavenumber (for the second-order central difference scheme) that was derived in Section 2.3. Thus, in the present case the corresponding λ in the model equation, $y' = \lambda y$, is $-ik'c$, which is purely imaginary. Thus, we would know, for example, that time advancement with the explicit Euler or second-order Runge–Kutta would lead to numerical instabilities. On the other hand if the leapfrog method is used, the maximum time step would be given by

$$\Delta t_{\max} = \frac{1}{k'c} = \frac{\Delta x}{c\ \sin(k\Delta x)}.$$

Again we will consider the worst case scenario, which leads to

$$\Delta t_{\max} = \frac{\Delta x}{c}$$

or

$$\frac{c\Delta t}{\Delta x} \leq 1. \tag{5.22}$$

The non-dimensional quantity $c\Delta t/\Delta x$ is called the CFL number, which is named after the mathematicians Courant, Friedrich, and Lewy. In numerical solutions of wave or convection type equations, the term "CFL number" is often used as an indicator of the stability of a numerical method. For example, if instead of leapfrog we had applied a fourth-order Runge–Kutta (in conjunction with the second-order finite difference for the spatial derivative) to (5.11), then in terms of the CFL number, the stability restriction would have been expressed as (see Figure 4.8)

$$\text{CFL} \leq 2.83. \tag{5.23}$$

One of the useful insights that can be deduced from the modified wavenumber analysis is the relationship between the maximum time step and the accuracy of the spatial differencing, which is used to discretize a partial differential equation. We have seen in examples of both the heat and convection equations, that the maximum time step allowed is limited by the *worst case scenario*, which is inversely proportional to the maximum value of the corresponding modified wavenumber. In Figure 2.2 the modified wavenumbers for three finite difference schemes were plotted. Note that the more accurate schemes have higher peak values for their modified wavenumbers. This means that in general, the more accurate spatial differencing schemes impose more restrictive constraints on the time step. This result is, of course, in accordance with our intuition; the more accurate schemes do a better job of resolving the high wavenumber components (small scales) of the solution, and the small scales have faster time scales that require smaller time steps to capture them.

EXAMPLE 5.3 Modified Wavenumber Stability Analysis

We will use the modified wavenumber analysis to determine the stability of the numerical methods in Examples 5.1 and 5.2. Applying a modified wavenumber analysis to the heat equation of Example 5.1 results in the following ordinary differential equation

$$\frac{d\psi}{dt} = -\alpha k'^2 \psi.$$

If the second-order spatial central differencing is used, the worst case (or the largest value) of k'^2 is

$$k'^2 = \frac{4}{\Delta x^2}.$$

Now using the stability limits we found in our treatment of ordinary differential equations we can predict the stability of various marching methods applied to this partial differential equation. For the application of the explicit Euler method we get a time-step constraint of

$$\Delta t \leq \frac{\Delta x^2}{2\alpha},$$

which is identical to that of the more general (and difficult) eigenvalue analysis. For the numerical values of Example 5.1 this constraint results in $\Delta t \leq 0.00125$. For fourth-order Runge–Kutta we predict that

$$\Delta t \leq \frac{2.79\Delta x^2}{4\alpha} = 0.00174$$

for stable solution. Since the modified wavenumber for this particular equation and the differencing scheme used is a negative real number, we would predict that *marching with leapfrog would result in an unstable solution.*

Similarly, we may analyze the stability of the numerical solution of the convection equation in Example 5.2. A modified wavenumber analysis of the equation yields

$$\frac{d\psi}{dt} = -ick'\psi.$$

For the second-order central differencing scheme, the worst case (i.e., the largest) modified wavenumber is

$$k' = \frac{1}{\Delta x}.$$

Since $-ick'$ is pure imaginary we know that the Euler method would be unstable. Similarly, the time-step advancement by fourth-order Runge–Kutta should be limited by (see Figure 4.8)

$$\Delta t \leq \frac{2.83\Delta x}{c}.$$

Taking $\Delta x = 0.01$ and $c = 1$ as in Example 5.2 gives $\Delta t \leq 0.028$. The time step used with leapfrog would be limited by

$$\Delta t \leq \frac{\Delta x}{c} = 0.01.$$

In summary, the modified wavenumber analysis offers a useful procedure for the stability analysis of time-dependent partial differential equations. It readily applies the results derived for ODEs to PDEs. The domain of applicability of the modified wavenumber analysis is nearly the same as that for the von Neumann analysis, i.e., linear, constant-coefficient PDEs with uniformly spaced spatial grid. The modified wavenumber analysis can be applied to problems where the space and time discretizations are clearly distinct, for example, if one uses a third-order Runge–Kutta scheme for time advancement and a second-order finite difference for spatial discretization. However, some numerical algorithms for PDEs are written such that the temporal and spatial discretizations are intermingled (see for example, Exercises 5 and 7(c) at the end of this chapter and the Du Fort–Frankel scheme (5.30) in Section 5.6). For such schemes the von Neumann analysis is still applicable, but the modified wavenumber analysis is not.

5.4 Implicit Time Advancement

We have established that semi-discretization of the heat equation leads to a stiff system of ODEs. We have also seen that for the heat equation, the stability limits for explicit schemes are too stringent. For these reasons implicit methods are preferred for parabolic equations. A popular implicit scheme is the trapezoidal method (introduced in Section 4.6 for ODEs), which is often referred to as the Crank–Nicolson method when applied to the heat equation,

$$\frac{\partial \phi}{\partial t} = \alpha \frac{\partial^2 \phi}{\partial x^2}. \tag{5.1}$$

Application of the trapezoidal method to (5.1) leads to

$$\frac{\phi_j^{(n+1)} - \phi_j^{(n)}}{\Delta t} = \frac{\alpha}{2}\left[\frac{\partial^2 \phi^{(n+1)}}{\partial x^2} + \frac{\partial^2 \phi^{(n)}}{\partial x^2}\right]_j \qquad j = 1, 2, 3, \ldots, N-1.$$

The subscript j refers to the spatial grid and the superscript n refers to the time step. Approximating the spatial derivatives with the second-order finite difference scheme on a uniform mesh yields

$$\phi_j^{(n+1)} - \phi_j^{(n)} = \frac{\alpha \Delta t}{2}\left[\frac{\phi_{j+1}^{(n+1)} - 2\phi_j^{(n+1)} + \phi_{j-1}^{(n+1)}}{\Delta x^2} + \frac{\phi_{j+1}^{(n)} - 2\phi_j^{(n)} + \phi_{j-1}^{(n)}}{\Delta x^2}\right].$$

Let $\beta = \alpha \Delta t / 2\Delta x^2$. Collecting the unknowns (terms with the superscript $(n+1)$) on the left-hand side results in the following tridiagonal system of equations:

$$-\beta \phi_{j+1}^{(n+1)} + (1 + 2\beta)\phi_j^{(n+1)} - \beta \phi_{j-1}^{(n+1)} = \beta \phi_{j+1}^{(n)} + (1 - 2\beta)\phi_j^{(n)} + \beta \phi_{j-1}^{(n)}.$$

Thus, at every time step a tridiagonal system of equations must be solved. The right-hand side of the system is computed using data from the current time step, n, and the solution at the next step, $n+1$, is obtained from the solution of the tridiagonal system. In general, *application of an implicit method to a partial differential equation requires solving a system of algebraic equations.* In one dimension, this does not cause any difficulty since the resulting matrix is a simple tridiagonal and requires on the order of N arithmetic operations to solve (see Appendix).

We can investigate the stability properties of this scheme using the von Neumann analysis or the equivalent modified wavenumber analysis. Recall that when applied to the model equation $y' = \lambda y$, the amplification factor for the trapezoidal method was (see Section 4.6)

$$\sigma = \frac{1 + \lambda \Delta t/2}{1 - \lambda \Delta t/2}.$$

Using the modified wavenumber analysis, the amplification factor for the trapezoidal method applied to the heat equation is obtained by substituting $-\alpha k'^2$ for λ in this equation. Here, k' is the modified wavenumber which was derived in (5.19):

$$k'^2 = \frac{2}{\Delta x^2}[1 - \cos(k \Delta x)].$$

Thus,

$$\sigma = \frac{1 - \frac{\alpha \Delta t}{\Delta x^2}[1 - \cos(k\Delta x)]}{1 + \frac{\alpha \Delta t}{\Delta x^2}[1 - \cos(k\Delta x)]}.$$

Since $1 - \cos(k\Delta x) \geq 0$, the denominator of σ is larger than its numerator, and hence $|\sigma| \leq 1$. Thus, we do not even have to identify the worst case scenario, the method is unconditionally stable.

Notice that for large $\alpha \Delta t/\Delta x^2$, σ approaches -1, which leads to temporal oscillations in the solution. However, the solution will always remain bounded. These undesirable oscillations in the solution are the basis for a controversial characteristic of the Crank–Nicolson method. To some, oscillation is an indicator of numerical inaccuracy and is interpreted as a warning: even though the method is stable, the time step is too large for accuracy and should be reduced. This warning feature is considered a desirable property. Others feel that it is more important to have smooth solutions (though possibly less accurate) because in more complex coupled problems (e.g., non-linear convection–diffusion) the oscillations can lead to further complications and inaccuracies.

A less accurate implicit method that does not lead to temporal oscillations at large time steps is the *backward Euler* method. Application of the backward

Euler time advancement and central space differencing to (5.1) results in

$$\phi_j^{(n+1)} - \phi_j^{(n)} = \alpha\Delta t\left[\frac{\phi_{j+1}^{(n+1)} - 2\phi_j^{(n+1)} + \phi_{j-1}^{(n+1)}}{\Delta x^2}\right].$$

Let $\gamma = \alpha\Delta t/\Delta x^2$. Collecting the unknowns on the left-hand side results in the following tridiagonal system of equations:

$$-\gamma\phi_{j+1}^{(n+1)} + (1+2\gamma)\phi_j^{(n+1)} - \gamma\phi_{j-1}^{(n+1)} = \gamma\phi_j^{(n)} \quad j = 1, 2, 3, \ldots, N-1.$$

Thus, the cost of applying the backward Euler scheme, which is only first-order accurate, is virtually the same as that for the second-order accurate Crank–Nicolson method. In both cases the major cost is in solving a tridiagonal system. Recall from Section 4.4 that the amplification factor for the backward Euler method when applied to $y' = \lambda y$ is

$$\sigma = \frac{1}{1 - \lambda\Delta t}.$$

Thus, for the heat equation, the amplification factor is

$$\sigma = \frac{1}{1 + 2\frac{\alpha\Delta t}{\Delta x^2}[1 - \cos(k\Delta x)]}.$$

The denominator is always larger than 1, and therefore, as expected, application of the backward Euler scheme to the heat equation is unconditionally stable. However, in contrast to the Crank–Nicolson scheme, $\sigma \longrightarrow 0$ as Δt becomes very large, and the solution does not exhibit undesirable oscillations (although it would be inaccurate).

EXAMPLE 5.4 Crank–Nicolson for the Heat Equation

We consider the same inhomogeneous heat equation as in Example 5.1. Taking $\beta = \alpha\Delta t/2\Delta x^2$, the tridiagonal system for the Crank–Nicolson time advancement of this equation is

$$\begin{aligned}-\beta T_{j+1}^{(n+1)} &+ (1+2\beta)T_j^{(n+1)} - \beta T_{j-1}^{(n+1)} \\ &= \beta T_{j+1}^{(n)} + (1-2\beta)T_j^{(n)} + \beta T_{j-1}^{(n)} + \Delta t\frac{f_j^{(n)} + f_j^{(n+1)}}{2},\end{aligned}$$

where, as before, f is the inhomogeneous term

$$f_j^{(n)} = (\pi^2 - 1)e^{-t_n}\sin\pi x_j.$$

Crank–Nicolson is unconditionally stable and we may therefore take a much larger time step than the $\Delta t = 0.001$ used in Example 5.1. Taking $\alpha = 1$ and $\Delta t = 0.05$, a very accurate solution to time $t = 2.0$ is calculated with only a fiftieth of the number of time steps taken in Example 5.1 (see Figure 5.5).

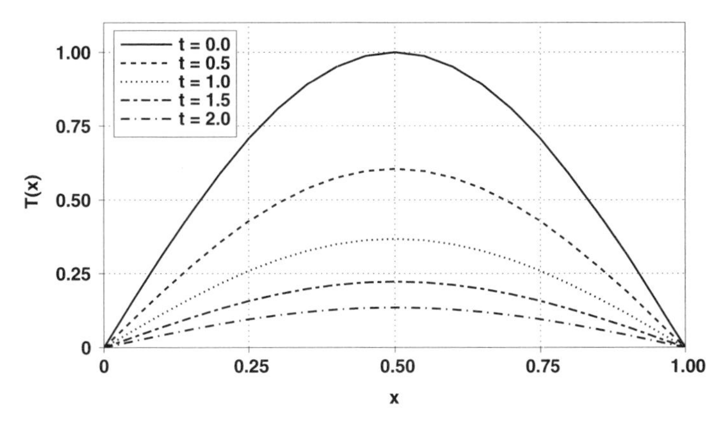

Figure 5.5 Numerical solution of the heat equation in Example 5.1 using the Crank–Nicolson method with $\Delta t = 0.05$.

The price paid for this huge decrease in the number of time steps is the cost of solving a tridiagonal system at each time step. However, algorithms for performing this task are very efficient (see Appendix), and in this example Crank–Nicolson offers a more efficient solution. This solution agrees to within a couple of percentage points with the exact solution. Larger time steps will give stable but less accurate solutions.

5.5 Accuracy via Modified Equation

We typically think of a numerical solution of a PDE as a set of numbers defined on a discrete set of space and time grid points. We can also think of the numerical solution as a continuous differentiable function that has the same values as the numerical solution on the computational grid points. In this section we will refer to this interpolant as the numerical solution. Since the numerical solution is an approximation to the exact solution, it does not *exactly* satisfy the continuous partial differential equation at hand, but it satisfies a *modified* equation. We shall derive the actual equation that a numerical solution satisfies and show how this knowledge can be used to select the numerical parameters of a method, resulting in better accuracy. In the next section we will show how this approach is used to identify an inconsistent numerical method.

Consider the heat equation (5.1). Let $\tilde{\phi}$ be the exact solution and ϕ be a continuous and differentiable function that assumes the same values as the numerical solution on the space–time grid. As an example, consider the discretization resulting from the application of the explicit Euler and second-order spatial differencing to (5.1):

$$\frac{\phi_j^{(n+1)} - \phi_j^{(n)}}{\Delta t} = \alpha \frac{\phi_{j+1}^{(n)} - 2\phi_j^{(n)} + \phi_{j-1}^{(n)}}{\Delta x^2}. \tag{5.24}$$

Let $L[\phi]$ be the difference operator:

$$L[\phi_j^{(n)}] = \frac{\phi_j^{(n+1)} - \phi_j^{(n)}}{\Delta t} - \alpha \frac{\phi_{j+1}^{(n)} - 2\phi_j^{(n)} + \phi_{j-1}^{(n)}}{\Delta x^2}. \tag{5.25}$$

Note that $L[\phi_j] = 0$ if ϕ satisfies (5.24). Given a function ϕ and a set of grid points in space and time, $L[\phi_j^{(n)}]$ is well defined. To obtain the modified equation, every term in (5.25) is expanded in Taylor series about $\phi_j^{(n)}$, and the resulting series are substituted in (5.25). For example,

$$\phi_j^{(n+1)} = \phi_j^{(n)} + \Delta t \frac{\partial \phi_j}{\partial t}^{(n)} + \frac{\Delta t^2}{2} \frac{\partial^2 \phi_j}{\partial t^2}^{(n)} + \cdots.$$

Thus,

$$\frac{\phi_j^{(n+1)} - \phi_j^{(n)}}{\Delta t} = \frac{\partial \phi_j}{\partial t}^{(n)} + \frac{\Delta t}{2} \frac{\partial^2 \phi_j}{\partial t^2}^{(n)} + \cdots.$$

Similarly,

$$\frac{\phi_{j+1}^{(n)} - 2\phi_j^{(n)} + \phi_{j-1}^{(n)}}{\Delta x^2} = \left.\frac{\partial^2 \phi^{(n)}}{\partial x^2}\right|_j + \left.\frac{\Delta x^2}{12} \frac{\partial^4 \phi^{(n)}}{\partial x^4}\right|_j + \cdots.$$

Substitution in (5.25) leads to

$$L[\phi_j^{(n)}] - \left(\frac{\partial \phi_j}{\partial t}^{(n)} - \alpha \left.\frac{\partial^2 \phi^{(n)}}{\partial x^2}\right|_j \right) = -\alpha \left.\frac{\Delta x^2}{12} \frac{\partial^4 \phi^{(n)}}{\partial x^4}\right|_j + \frac{\Delta t}{2} \frac{\partial^2 \phi_j^{(n)}}{\partial t^2} + \cdots. \tag{5.26}$$

This equation was derived without reference to a specific set of space–time grid points. In other words, the indices j and n are generic, and equation (5.26) applies to any point in space and time. That is,

$$L[\phi] - \left(\frac{\partial \phi}{\partial t} - \alpha \frac{\partial^2 \phi}{\partial x^2} \right) = -\alpha \frac{\Delta x^2}{12} \frac{\partial^4 \phi}{\partial x^4} + \frac{\Delta t}{2} \frac{\partial^2 \phi}{\partial t^2} + \cdots \tag{5.27}$$

Let ϕ be the solution of the discrete equation (5.24). Then, $L[\phi] = 0$, and it can be seen that the numerical solution actually satisfies the following *modified* differential equation

$$\frac{\partial \phi}{\partial t} - \alpha \frac{\partial^2 \phi}{\partial x^2} = \alpha \frac{\Delta x^2}{12} \frac{\partial^4 \phi}{\partial x^4} - \frac{\Delta t}{2} \frac{\partial^2 \phi}{\partial t^2} + \cdots$$

instead of (5.1). Note that as Δt and Δx approach zero, the modified equation approaches the exact PDE. The modified equation also shows that the numerical method is first-order accurate in time and second-order in space. Furthermore, if either the time step or the spatial mesh size is reduced without reducing the other, one simply gets to the point of diminishing returns, as the overall error remains finite. However, there may be a possibility of cancelling errors by a judicious choice of the time step in terms of the spatial step. We shall explore this possibility next.

If $\tilde{\phi}$ is the exact solution of the PDE in (5.1), then

$$\frac{\partial \tilde{\phi}}{\partial t} = \alpha \frac{\partial^2 \tilde{\phi}}{\partial x^2} \tag{5.28}$$

and

$$L[\tilde{\phi}] = \epsilon \neq 0,$$

where

$$\epsilon = -\alpha \frac{\Delta x^2}{12} \frac{\partial^4 \tilde{\phi}}{\partial x^4} + \frac{\Delta t}{2} \frac{\partial^2 \tilde{\phi}}{\partial t^2} + \cdots.$$

But, since $\tilde{\phi}$ satisfies (5.28), we have

$$\frac{\partial^2 \tilde{\phi}}{\partial t^2} = \alpha \frac{\partial^2 \tilde{\phi}}{\partial t \partial x^2} = \alpha^2 \frac{\partial^4 \tilde{\phi}}{\partial x^4}.$$

Therefore,

$$\epsilon = \left(-\alpha \frac{\Delta x^2}{12} + \alpha^2 \frac{\Delta t}{2} \right) \frac{\partial^4 \tilde{\phi}}{\partial x^4} + \cdots.$$

Thus, we can increase the accuracy of the numerical solution by setting the term inside the parenthesis to zero, i.e.,

$$\alpha \frac{\Delta x^2}{12} = \alpha^2 \frac{\Delta t}{2}.$$

In other words, by selecting the space and time increments such that

$$\frac{\alpha \Delta t}{\Delta x^2} = \frac{1}{6},$$

we could significantly increase the accuracy of the method. This constraint is within the stability limit derived earlier (i.e., $\alpha \Delta t / \Delta x^2 \leq 1/2$), but is rather restrictive, requiring a factor of 3 reduction in time step from the stability limit which is rather stiff to begin with.

5.6 Du Fort–Frankel Method: An Inconsistent Scheme

An interesting application of the modified equation analysis arises in the study of a numerical scheme developed by Du Fort and Frankel for the solution of the heat equation. We will first derive the method and then analyze it using its modified equation. The method is derived in two steps. Consider the combination of the leapfrog time advancement (Section 4.9) and the second-order central spatial differencing

$$\frac{\phi_j^{(n+1)} - \phi_j^{(n-1)}}{2\Delta t} = \frac{\alpha}{\Delta x^2}\left[\phi_{j+1}^{(n)} - 2\phi_j^{(n)} + \phi_{j-1}^{(n)}\right] + O(\Delta t^2, \Delta x^2). \qquad (5.29)$$

This scheme is formally second-order accurate in both space and time. However, it is unconditionally unstable (see Example 5.3). The Du Fort–Frankel scheme is obtained by substituting for $\phi_j^{(n)}$, in the right-hand side of (5.29), the following second-order approximation

$$\phi_j^{(n)} = \frac{\phi_j^{(n+1)} + \phi_j^{(n-1)}}{2} + O(\Delta t^2).$$

Rearranging terms results in

$$(1 + 2\gamma)\phi_j^{(n+1)} = (1 - 2\gamma)\phi_j^{(n-1)} + 2\gamma\phi_{j+1}^{(n)} + 2\gamma\phi_{j-1}^{(n)}, \qquad (5.30)$$

where $\gamma = \alpha\Delta t/\Delta x^2$. It turns out that this method is *unconditionally stable*! In other words, the Du Fort–Frankel scheme has the same stability property as for implicit methods, but with a lot less work per time step. Recall that application of an implicit method requires matrix inversions at each time step, whereas this method does not. As we shall see, this is too good to be true.

Let us derive the modified equation for the Du Fort–Frankel scheme. Substituting Taylor series expansions for $\phi_{j+1}^{(n)}, \phi_{j-1}^{(n)}, \phi_j^{(n+1)}$, and $\phi_j^{(n-1)}$ into (5.30) and performing some algebra leads to

$$\frac{\partial \phi}{\partial t} - \alpha\frac{\partial^2 \phi}{\partial x^2} = -\frac{\Delta t^2}{6}\frac{\partial^3 \phi}{\partial t^3} + \frac{\alpha \Delta x^2}{12}\frac{\partial^4 \phi}{\partial x^4} - \frac{\alpha \Delta t^2}{\Delta x^2}\frac{\partial^2 \phi}{\partial t^2} - \frac{\alpha \Delta t^4}{12\Delta x^2}\frac{\partial^4 \phi}{\partial t^4} + \cdots.$$

This is the modified equation for the Du Fort–Frankel scheme for the heat equation. It reveals a fundamental problem on the right-hand side. The difficulty is due to the third and some of the subsequent terms on the right-hand side. For a given time step, if we refine the spatial mesh, the error actually increases! Thus, one cannot increase the accuracy of the numerical solution by arbitrarily letting $\Delta x \to 0$ and $\Delta t \to 0$. For example, the third term approaches zero only if Δt approaches zero faster than Δx does. For this reason the Du Fort–Frankel scheme is considered to be an inconsistent numerical method.

EXAMPLE 5.5 Du Fort–Frankel

Again considering the heat equation of Example 5.1 and taking $\gamma = \alpha\Delta t/\Delta x^2$ the advancement algorithm for Du Fort–Frankel is

$$(1 + 2\gamma)T_j^{(n+1)} = 2\gamma T_{j+1}^{(n)} + (1 - 2\gamma)T_j^{(n-1)} + 2\gamma T_{j-1}^{(n)} + 2\Delta t f_j^{(n)},$$

where f is the inhomogeneous term,

$$f_j^{(n)} = (\pi^2 - 1)e^{-t_n}\sin \pi x_j.$$

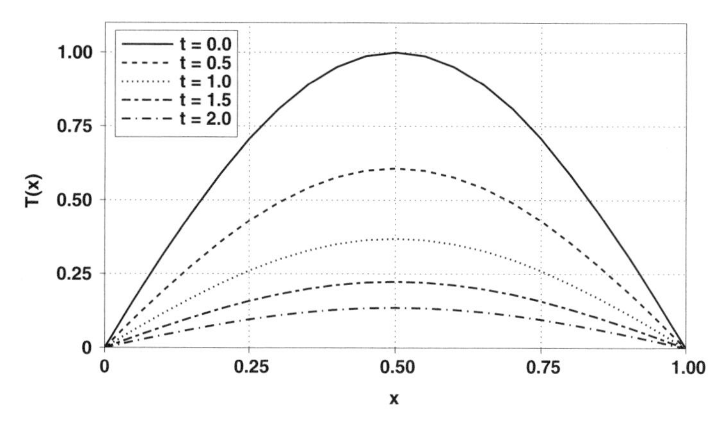

Figure 5.6 Numerical solution of the heat equation in Example 5.1 using the Du Fort–Frankel method with $\Delta t = 0.025$, $\Delta x = 0.05$.

Taking $\alpha = 1$ and $\Delta t = 0.025$, we repeat the calculation of Example 5.4 using the Du Fort–Frankel time advancement. This solution has comparable accuracy to the Crank–Nicolson method with twice the number of time steps (see Figure 5.6).

Like Crank–Nicolson, the Du Fort–Frankel scheme is unconditionally stable, but has the advantage of being of explicit form, so matrix inversions are not necessary to advance the solution and it is therefore simpler to program and cheaper to solve (on a per time-step basis). However, this section shows that the method is inconsistent. With larger choices of Δt with respect to Δx, the coefficients of some of the error terms in the modified equation are no longer small and one actually solves a different partial differential equation. For example, taking $\Delta t = 2\Delta x = 0.1$ the solution is stable but grossly incorrect (resulting in negative temperatures!) as shown in Figure 5.7.

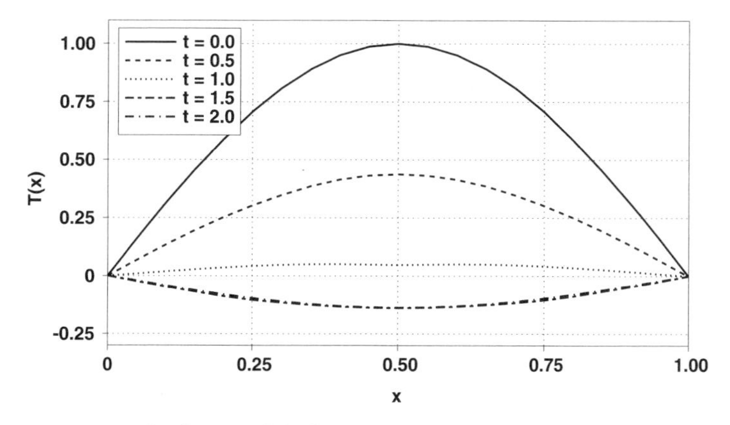

Figure 5.7 Numerical solution of the heat equation in Example 5.1 using the Du Fort–Frankel method with $\Delta t = 0.1$, $\Delta x = 0.05$.

5.7 Multi-Dimensions

Up to this point we have considered partial differential equations in one space dimension and time. Most physical problems are posed in two- or three-dimensional space. In this and the following sections we will explore some of the main issues and algorithms for solving partial differential equations in multi-dimensional space and time. We will see that as far as implementation of a numerical scheme is concerned, higher dimensions do not cause additional complications, as long as *explicit time advancement* is used. However, straightforward applications of implicit schemes lead to large systems of equations that can easily overwhelm computer memory requirements. In Section 5.9 we will introduce a clever algorithm to circumvent this problem.

Consider the two-dimensional heat equation

$$\frac{\partial \phi}{\partial t} = \alpha \left(\frac{\partial^2 \phi}{\partial x^2} + \frac{\partial^2 \phi}{\partial y^2} \right), \tag{5.31}$$

with ϕ prescribed on the boundaries of a rectangular domain. For numerical solution, we first introduce a grid in the xy plane as in Figure 5.8. Let $\phi_{l,j}^{(n)}$ denote the value of ϕ at the grid point (l, j) at time step n. We use $M + 1$ grid points in x and $N + 1$ points in y. The boundary points are at $l = 0, M$ and $j = 0, N$.

Application of any *explicit* numerical method is very straightforward. For example, consider the explicit Euler in conjunction with the second-order central

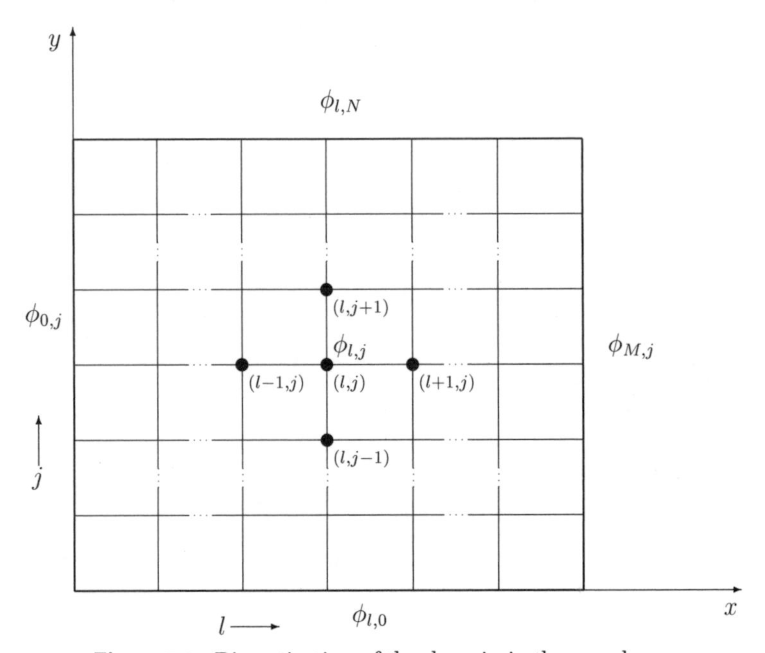

Figure 5.8 Discretization of the domain in the xy plane.

finite difference approximation for the spatial derivatives

$$\frac{\phi_{l,j}^{(n+1)} - \phi_{l,j}^{(n)}}{\Delta t} = \alpha \left[\frac{\phi_{l+1,j}^{(n)} - 2\phi_{l,j}^{(n)} + \phi_{l-1,j}^{(n)}}{\Delta x^2} + \frac{\phi_{l,j+1}^{(n)} - 2\phi_{l,j}^{(n)} + \phi_{l,j-1}^{(n)}}{\Delta y^2} \right]$$

$$l = 1, 2, \ldots, M-1 \quad j = 1, 2, \ldots, N-1, \quad n = 0, 1, 2, \cdots. \tag{5.32}$$

Given an initial condition on the grid points, denoted by $\phi_{l,j}^{(0)}$, for each l and j one simply marches forward in time to obtain the solution at the subsequent time steps. When $l = 1$ or $l = M - 1$, or $j = 1$ or $j = N - 1$, boundary values are required, and their values from the prescribed (in this case Dirichlet) boundary conditions are used. For example, for $n = 0$, all the terms with superscript 0 are obtained from the initial condition; equation (5.32) is then used to calculate $\phi_{l,j}^{(1)}$ for all the interior points. Next, $\phi_{l,j}^{(1)}$ is used to compute $\phi_{l,j}^{(2)}$, and so on. Note that boundary conditions can be functions of time. Thus, at $t = n\Delta t$, the prescribed boundary data, $\phi_{l,N}^{(n)}$, for example, are used when needed.

The stability properties of this scheme can be analyzed in the same manner as in the one-dimensional case. Considering solutions of the form $\phi = \psi(t)e^{ik_1x+ik_2y}$, the semi-discretized version of (5.31) transforms to

$$\frac{d\psi}{dt} = -\alpha\left(k_1'^2 + k_2'^2\right)\psi \tag{5.33}$$

where, k_1' and k_2' are the modified wavenumbers corresponding to x and y directions respectively:

$$\begin{aligned} k_1'^2 &= \frac{2}{\Delta x^2}[1 - \cos(k_1 \Delta x)] \\ k_2'^2 &= \frac{2}{\Delta y^2}[1 - \cos(k_2 \Delta y)]. \end{aligned} \tag{5.34}$$

Since $-\alpha(k_1'^2 + k_2'^2)$ is real and negative and we are using the explicit Euler time advancement, for stability we must have

$$\Delta t \le \frac{2}{\alpha\left[\frac{2}{\Delta x^2}[1 - \cos(k_1\Delta x)] + \frac{2}{\Delta y^2}[1 - \cos(k_2\Delta y)]\right]}.$$

The worst case is when $\cos(k_1\Delta x) = -1$ and $\cos(k_2\Delta y) = -1$. Thus,

$$\Delta t \le \frac{1}{2\alpha\left(\frac{1}{\Delta x^2} + \frac{1}{\Delta y^2}\right)}. \tag{5.35}$$

This is the basic stability criterion for the heat equation in two dimensions. It is the stability limit for the numerical method consisting of the explicit Euler time advancement and second-order central differencing for spatial derivatives. As in Section 5.3, we can readily obtain the stability limits for different time advancement schemes or spatial differencing using the modified wavenumber analysis. In the special case, $\Delta x = \Delta y = h$, we obtain

$$\Delta t \leq \frac{h^2}{4\alpha}, \tag{5.36}$$

which is two times more restrictive than the one-dimensional case. Similarly, in three-dimensions one obtains

$$\Delta t \leq \frac{h^2}{6\alpha}. \tag{5.37}$$

5.8 Implicit Methods in Higher Dimensions

As in the case of the one-dimensional heat equation, the predicament of severe time-step restriction with explicit schemes suggests using implicit methods. In addition, we have shown in Section 5.7 that the stability restriction in multi-dimensional problems is more severe than that in one dimension. Thus, we are very motivated to explore the possibility of using implicit methods for multi-dimensional problems.

As an example, consider application of the Crank–Nicolson scheme to the two-dimensional heat equation:

$$\frac{\phi^{(n+1)} - \phi^{(n)}}{\Delta t} = \frac{\alpha}{2}\left[\frac{\partial^2\phi^{(n+1)}}{\partial x^2} + \frac{\partial^2\phi^{(n+1)}}{\partial y^2} + \frac{\partial^2\phi^{(n)}}{\partial x^2} + \frac{\partial^2\phi^{(n)}}{\partial y^2}\right]. \tag{5.38}$$

Using second-order finite differences in space and assuming $\Delta x = \Delta y = h$, we obtain:

$$\begin{aligned}\phi_{l,j}^{(n+1)} - \phi_{l,j}^{(n)} = {} & \frac{\alpha\Delta t}{2h^2}\left[\phi_{l+1,j}^{(n+1)} - 2\phi_{l,j}^{(n+1)} + \phi_{l-1,j}^{(n+1)} + \phi_{l,j+1}^{(n+1)} - 2\phi_{l,j}^{(n+1)} + \phi_{l,j-1}^{(n+1)}\right] \\ & + \frac{\alpha\Delta t}{2h^2}\left[\phi_{l+1,j}^{(n)} - 2\phi_{l,j}^{(n)} + \phi_{l-1,j}^{(n)} + \phi_{l,j+1}^{(n)} - 2\phi_{l,j}^{(n)} + \phi_{l,j-1}^{(n)}\right].\end{aligned} \tag{5.39}$$

Let $\beta = \alpha\Delta t/2h^2$, collecting the unknowns on the left-hand side yields

$$\begin{aligned}-\beta\phi_{l+1,j}^{(n+1)} + (1+4\beta)\phi_{l,j}^{(n+1)} - \beta\phi_{l-1,j}^{(n+1)} - \beta\phi_{l,j+1}^{(n+1)} - \beta\phi_{l,j-1}^{(n+1)} \\ = \beta\phi_{l+1,j}^{(n)} + (1-4\beta)\phi_{l,j}^{(n)} + \beta\phi_{l-1,j}^{(n)} + \beta\phi_{l,j+1}^{(n)} + \beta\phi_{l,j-1}^{(n)}.\end{aligned} \tag{5.40}$$

This is a gigantic system of algebraic equations for $\phi_{l,j}^{(n+1)}$, ($l = 1, 2, \ldots, M-1$; $j = 1, 2, \ldots, N-1$).

The best way to see the form of the matrix and gain an appreciation for the problem at hand is to write down a few of the equations. We will first order the elements of the unknown vector ϕ as follows

$$\phi = \begin{bmatrix} \phi_{1,1} \\ \phi_{2,1} \\ \phi_{3,1} \\ \vdots \\ \phi_{M-1,1} \\ \phi_{1,2} \\ \phi_{2,2} \\ \phi_{3,2} \\ \vdots \\ \phi_{M-1,2} \\ \vdots \\ \vdots \\ \phi_{1,N-1} \\ \phi_{2,N-1} \\ \phi_{3,N-1} \\ \vdots \\ \phi_{M-1,N-1} \end{bmatrix}^{(n+1)} \tag{5.41}$$

Note that ϕ is a vector with $(M-1) \times (N-1)$ unknown elements corresponding to the number of interior grid points in the domain. Now, let us write down some of the algebraic equations. For $l = 1$ and $j = 1$, equation (5.40) becomes

$$-\beta\phi_{2,1}^{(n+1)} + (1+4\beta)\phi_{1,1}^{(n+1)} - \beta\phi_{0,1}^{(n+1)} - \beta\phi_{1,2}^{(n+1)} - \beta\phi_{1,0}^{(n+1)} = F_{1,1}^{(n)}, \tag{5.42}$$

where $F_{1,1}^{(n)}$ is the right-hand side of equation (5.40), which is known because every term in it is evaluated at time step n. Next, we note that $\phi_{0,1}^{(n+1)}$ and $\phi_{1,0}^{(n+1)}$ in (5.42) are known from the boundary conditions and therefore should be moved to the right-hand side of (5.42). Thus, the equation corresponding to $l = 1,\ j = 1$ becomes

$$-\beta\phi_{2,1}^{(n+1)} + (1+4\beta)\phi_{1,1}^{(n+1)} - \beta\phi_{1,2}^{(n+1)} = F_{1,1}^{(n)} + \beta\phi_{0,1}^{(n+1)} + \beta\phi_{1,0}^{(n+1)}.$$

The next equation in the ordering of ϕ shown in (5.41) is obtained by letting $l = 2,\ j = 1$ in (5.40). Again, after moving the boundary term to the right-hand side we get

$$-\beta\phi_{3,1}^{(n+1)} + (1+4\beta)\phi_{2,1}^{(n+1)} - \beta\phi_{1,1}^{(n+1)} - \beta\phi_{2,2}^{(n+1)} = F_{21}^{(n)} + \beta\phi_{2,0}^{(n+1)}.$$

This process is continued for all the remaining $l = 3, 4, \ldots, (M-1)$ and $j = 1$. Next, j is set equal to 2 and all the equations in (5.40) corresponding to $l = 1, 2, 3, \ldots, (M-1)$ are accounted for. The process continues until $j = (N-1)$.

After writing a few of these equations in matrix form, we see that a pattern emerges. The resulting $[(M-1)\times(N-1)]\times[(M-1)\times(N-1)]$ matrix is of *block-tridiagonal* form

$$\mathcal{A}=\begin{bmatrix} B & C & & & \\ A & B & C & & \\ & & \ddots & \ddots & \ddots \\ & & & A & B \end{bmatrix}, \tag{5.43}$$

where A, B, and C are $(M-1)\times(M-1)$ matrices, and there are N such B matrices on the diagonal. In the present case, A and C are diagonal matrices whereas B is tridiagonal,

$$B=\begin{bmatrix} 1+4\beta & -\beta & & \\ -\beta & 1+4\beta & -\beta & \\ & \ddots & \ddots & \ddots \\ & & -\beta & 1+4\beta \end{bmatrix} \qquad A,C=\begin{bmatrix} -\beta & & & \\ & -\beta & & \\ & & \ddots & \\ & & & -\beta \end{bmatrix}.$$

Clearly, $\mathcal{A}$ is very large. For example, for $M=101$ and $N=101$, $\mathcal{A}$ has 10^8 elements. However, $\mathcal{A}$ is banded, and there is no need to store the zero elements of the matrix outside its central band of width $2M$; in this case the required memory is reduced to $2(M-1)^2(N-1)$. For the present case of uniform mesh spacings in both x and y directions, there are other tricks that can be used to reduce the required memory even further (one such method is described in Chapter 6). However, for now, we are not going to discuss these options further and opt instead for an alternative approach that is also applicable to higher dimensional problems and has more general applicability, including to differential equations with non-constant coefficients and non-uniform mesh distributions.

5.9 Approximate Factorization

The difficulty of working with large matrices resulting from straightforward implementation of implicit schemes to PDEs in higher dimensions has led to the development of the so-called split or factored schemes. As the name implies, such schemes split a multi-dimensional problem to a series of one-dimensional ones, which are much easier to solve. Of course, in general, this conversion cannot be done exactly and some error is incurred. However, as we will show below, the splitting error is of the same order as the error already incurred in discretizing the problem in space and time. That is, the splitting approximation does not erode the order of accuracy of the scheme. This is the second time that we use this clever "trick" of numerical analysis; the first time was in the implicit solution of non-linear ordinary differential equations by linearization.

In the case of interest here, we note that the large matrix in (5.43) is obtained after making a numerical approximation to the two-dimensional heat equation by the Crank–Nicolson scheme. Therefore, one is not obligated to solve an approximate system of equations exactly. It suffices to obtain the solution to within the error already incurred by the spatial and temporal discretizations. Thus, we are going to circumvent large matrices while maintaining the same order of accuracy.

Consider application of the Crank–Nicolson method and the second-order spatial differencing to the two-dimensional heat equation (with homogeneous Dirichlet boundary conditions). Let's rewrite equation (5.39) in the operator notation

$$\frac{\phi^{(n+1)} - \phi^{(n)}}{\Delta t} = \frac{\alpha}{2} A_x\left[\phi^{(n+1)} + \phi^{(n)}\right] + \frac{\alpha}{2} A_y\left[\phi^{(n+1)} + \phi^{(n)}\right] + O(\Delta t^2) + O(\Delta x^2) + O(\Delta y^2), \tag{5.44}$$

where A_x and A_y are the difference operators representing the spatial derivatives in x and y directions respectively. For example, $A_x \phi$ is a vector of length $(N-1) \times (M-1)$ with elements defined as

$$\frac{\phi_{i+1,j} - 2\phi_{i,j} + \phi_{i-1,j}}{\Delta x^2} \quad i = 1, 2, \ldots, M-1 \quad j = 1, 2, \ldots, N-1.$$

We are also keeping track of errors to ensure that any further approximations that are going to be made will be within the order of these original errors. Equation (5.44) can be recast in the following form:

$$\left[I - \frac{\alpha \Delta t}{2} A_x - \frac{\alpha \Delta t}{2} A_y\right] \phi^{(n+1)} = \left[I + \frac{\alpha \Delta t}{2} A_x + \frac{\alpha \Delta t}{2} A_y\right] \phi^{(n)} + \Delta t\left[O(\Delta t^2) + O(\Delta x^2) + O(\Delta y^2)\right].$$

Each side can be rearranged into a partial factored form as follows:

$$\left(I - \frac{\alpha \Delta t}{2} A_x\right)\left(I - \frac{\alpha \Delta t}{2} A_y\right) \phi^{(n+1)} - \frac{\alpha^2 \Delta t^2}{4} A_x A_y \phi^{(n+1)}$$

$$= \left(I + \frac{\alpha \Delta t}{2} A_x\right)\left(I + \frac{\alpha \Delta t}{2} A_y\right) \phi^{(n)} - \frac{\alpha^2 \Delta t^2}{4} A_x A_y \phi^{(n)} + \Delta t\left[O(\Delta t^2) + O(\Delta x^2) + O(\Delta y^2)\right].$$

Taking the "cross terms" to the right-hand side and combining them leads to

$$\left(I - \frac{\alpha\Delta t}{2}A_x\right)\left(I - \frac{\alpha\Delta t}{2}A_y\right)\phi^{(n+1)} = \left(I + \frac{\alpha\Delta t}{2}A_x\right)\left(I + \frac{\alpha\Delta t}{2}A_y\right)\phi^{(n)}$$

$$+ \frac{\alpha^2\Delta t^2}{4}A_x A_y(\phi^{(n+1)} - \phi^{(n)}) + \Delta t\big[O(\Delta t^2) + O(\Delta x^2) + O(\Delta y^2)\big].$$

Using Taylor series in t, it is easy to see that, $\phi^{(n+1)} - \phi^{(n)} = O(\Delta t)$. Thus, as with the overall error of the scheme, the cross terms are $O(\Delta t^3)$ and can be neglected without any loss in the order of accuracy. Hence, we arrive at the factored form of the discrete equations

$$\left(I - \frac{\alpha\Delta t}{2}A_x\right)\left(I - \frac{\alpha\Delta t}{2}A_y\right)\phi^{(n+1)} = \left(I + \frac{\alpha\Delta t}{2}A_x\right)\left(I + \frac{\alpha\Delta t}{2}A_y\right)\phi^{(n)}. \tag{5.45}$$

This equation is much easier and more cost effective to implement than the large system encountered in the non-factored form. Basically, the multi-dimensional problem is reduced to a series of one-dimensional problems.

This is how the factored algorithm works. It is implemented in two steps. Let the (known) right-hand side of (5.45) be denoted by $\boldsymbol{f}$, and let

$$z = \left(I - \frac{\alpha\Delta t}{2}A_y\right)\phi^{(n+1)}. \tag{5.46}$$

Then, z can be obtained from the following equation, which is obtained directly from (5.45):

$$\left(I - \frac{\alpha\Delta t}{2}A_x\right)z = \boldsymbol{f}.$$

This equation can be recast into index notation

$$z_{i,j} - \left(\frac{\alpha\Delta t}{2}\right)\frac{z_{i-1,j} - 2z_{i,j} + z_{i+1,j}}{\Delta x^2} = f_{i,j}^n$$

or

$$-\frac{\alpha\Delta t}{2\Delta x^2}z_{i+1,j} + \left(1 + \frac{\alpha\Delta t}{\Delta x^2}\right)z_{i,j} - \frac{\alpha\Delta t}{2\Delta x^2}z_{i-1,j} = f_{i,j}. \tag{5.47}$$

Thus, for each $j = 1, 2, \ldots, N-1$, a simple tridiagonal system is solved for $z_{i,j}$. In the computer program that deals with this part of the problem, the tridiagonal solver is called within a simple loop running over the index j.

After calculating z, we obtain $\phi^{(n+1)}$ from (5.46):

$$\left(I - \frac{\alpha\Delta t}{2}A_y\right)\phi^{(n+1)} = z.$$

In index notation, we have

$$-\frac{\alpha\Delta t}{2\Delta y^2}\phi_{i,j+1}^{(n+1)} + \left(1+\frac{\alpha\Delta t}{\Delta y^2}\right)\phi_{i,j}^{(n+1)} - \frac{\alpha\Delta t}{2\Delta y^2}\phi_{i,j-1}^{(n+1)} = z_{i,j}. \tag{5.48}$$

For each $i = 1, 2, \ldots, M-1$, a tridiagonal system of equations is solved for $\phi_{i,j}^{(n+1)}$. This part is implemented in the computer program in an identical fashion to that used to solve for z, except that the loop is now over the index i.

Thus, with the factored algorithm, instead of solving one large system of size $(M-1)^2 \times (N-1)^2$, one solves $(M-1)$ tridiagonal systems of size $(N-1)$ and $(N-1)$ tridiagonal systems of size $(M-1)$. The number of arithmetic operations is on the order of MN, and the memory requirement is virtually negligible.

There is an important point that needs to be addressed with regard to the solution of the system (5.47). When $i = 1$ or N, boundary values for z are required in the form of $z_{0,j}$ or $z_{N,j}$. However, boundary conditions are only prescribed for ϕ, the original unknown in the heat equation. We can obtain the required boundary conditions for z from (5.46), the equation defining z. For example, at the $x = 0$ boundary, $z_{0,j}$ is computed from

$$z_{0,j} = \phi_{0,j}^{(n+1)} - \frac{\alpha\Delta t}{2}\frac{\phi_{0,j+1}^{(n+1)} - 2\phi_{0,j}^{(n+1)} + \phi_{0,j-1}^{(n+1)}}{\Delta y^2} \qquad j = 1, 2, \ldots, N-1.$$

Note that $\phi_{0,j}^{(n+1)}$'s are prescribed as (time dependent) Dirichlet boundary conditions for the heat equation. Similarly, boundary values of z can be obtained at the other boundary, x_N. If for example, $\phi(x = 0, y, t)$ is not a function of y along the left $(x = 0)$ boundary, then z would be equal to ϕ at the boundary. But, if the prescribed boundary condition happens to be a function of y, then z at the boundary differs from ϕ by an $O(\Delta t)$ correction proportional to the second derivative of ϕ on the boundary.

In three dimensions, the use of approximate factorization becomes an essential necessity. Straightforward application of implicit methods without splitting or factorization in three dimensions is virtually impossible. Fortunately, the extension of the approximate factorization scheme described in this section to three dimensions is trivial. The factored form of the Crank–Nicolson algorithm applied to the 3D heat equation is

$$\left(I - \frac{\alpha\Delta t}{2}A_x\right)\left(I - \frac{\alpha\Delta t}{2}A_y\right)\left(I - \frac{\alpha\Delta t}{2}A_z\right)\phi^{n+1}$$
$$= \left(I + \frac{\alpha\Delta t}{2}A_x\right)\left(I + \frac{\alpha\Delta t}{2}A_y\right)\left(I + \frac{\alpha\Delta t}{2}A_z\right)\phi^{n} \tag{5.49}$$

which is second order in space and time. The scheme can be implemented in the same manner as in 2D by introducing suitable intermediate variables with the corresponding boundary conditions.

EXAMPLE 5.6 Approximate Factorization for the Heat Equation

Consider the following inhomogeneous two-dimensional heat equation

$$\frac{\partial \phi}{\partial t} = \left(\frac{\partial^2 \phi}{\partial x^2} + \frac{\partial^2 \phi}{\partial y^2}\right) + q(x, y),$$

with homogeneous initial and boundary conditions

$$\phi(x, y, 0) = 0 \quad \phi(\pm 1, y, t) = 0 \quad \phi(x, \pm 1, t) = 0$$

and

$$q(x, y) = 2(2 - x^2 - y^2).$$

Suppose, we wish to integrate this equation to the steady state (i.e., to the point where $\partial\phi/\partial t = 0$). In fact, if the steady state solution is the only thing we are interested in, then the accuracy of the transient part of the solution is not important, and we can take large time steps to decrease the cost of the solution. An implicit method is therefore desirable. We choose the Crank–Nicolson scheme and use an approximate factorization to avoid solving a large system. The source term q is not a function of time and therefore $\boldsymbol{q}^{(n+1)} = \boldsymbol{q}^{(n)}$ and the factorized system for advancing in time is (with $\alpha = 1$)

$$\left(I - \frac{\Delta t}{2} A_x\right)\left(I - \frac{\Delta t}{2} A_y\right)\boldsymbol{\phi}^{(n+1)} = \left(I + \frac{\Delta t}{2} A_x\right)\left(I + \frac{\Delta t}{2} A_y\right)\boldsymbol{\phi}^{(n)} + \Delta t \boldsymbol{q}.$$

The solution proceeds as follows. The right-hand side consists of known terms and therefore may be evaluated explicitly in steps. Taking

$$\boldsymbol{\xi}^{(n)} = \left(I + \frac{\Delta t}{2} A_y\right)\boldsymbol{\phi}^{(n)},$$

we may evaluate $\boldsymbol{\xi}^{(n)}$ at all points (i, j) by

$$\xi_{i,j}^{(n)} = \phi_{i,j}^{(n)} + \frac{\Delta t}{2\Delta y^2}\left(\phi_{i,j+1}^{(n)} - 2\phi_{i,j}^{(n)} + \phi_{i,j-1}^{(n)}\right).$$

Then, taking

$$\boldsymbol{r}^{(n)} = \left(I + \frac{\Delta t}{2} A_x\right)\boldsymbol{\xi}^{(n)} + \Delta t \boldsymbol{q},$$

the right-hand side $\boldsymbol{r}$ is calculated by

$$r_{i,j}^{(n)} = \xi_{i,j}^{(n)} + \frac{\Delta t}{2\Delta y^2}\left(\xi_{i+1,j}^{(n)} - 2\xi_{i,j}^{(n)} + \xi_{i-1,j}^{(n)}\right) + \Delta t q_{i,j}.$$

We are left with the following set of equations to solve for ϕ at the next time level $(n + 1)$:

$$\left(I - \frac{\Delta t}{2} A_x\right)\left(I - \frac{\Delta t}{2} A_y\right)\boldsymbol{\phi}^{(n+1)} = \boldsymbol{r}^{(n)}.$$

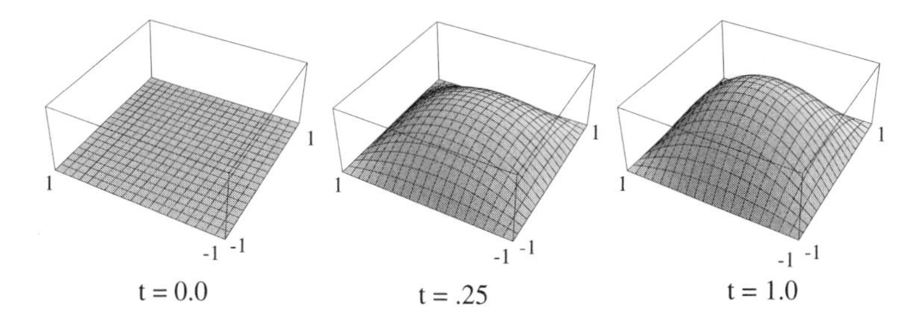

Figure 5.9 Numerical solution of 2D heat equation using the approximate factorization technique with $\Delta t = 0.05$ and $M = N = 20$. The solution at $t = 1$ is near steady state.

This is solved in two phases as outlined in the text. First we define

$$\eta^{(n+1)} = \left(I - \frac{\Delta t}{2} A_y\right)\phi^{(n+1)}$$

and solve the tridiagonal systems

$$\eta_{i,j}^{(n+1)} - \frac{\Delta t}{2\Delta x^2}\left(\eta_{i+1,j}^{(n+1)} - 2\eta_{i,j}^{(n+1)} + \eta_{i-1,j}^{(n+1)}\right) = r_{i,j} \quad i = 1, 2, \ldots, M-1,$$

for $j = 1, 2, \ldots, N-1$. Boundary conditions are needed for η, and for this problem, they are simply $\eta_{0,j} = \eta_{M,j} = 0$. Then using the definition of $\eta^{(n+1)}$ we solve $M-1$ tridiagonal systems to calculate $\phi^{(n+1)}$

$$\phi_{i,j}^{(n+1)} - \frac{\Delta t}{2\Delta y^2}\left(\phi_{i,j+1}^{(n+1)} - 2\phi_{i,j}^{(n+1)} + \phi_{i,j-1}^{(n+1)}\right) = \eta_{i,j}^{(n+1)} \quad j = 1, 2, \ldots, N-1,$$

for $i = 1, 2, \ldots, M-1$. Boundary conditions ($\phi_{i,0} = \phi_{i,N} = 0$) are applied to ϕ and we have obtained the solution ϕ at the time level $(n+1)$. The first set of numerical parameters chosen are $\Delta t = 0.05$ and $M = N = 20$, for which the results are plotted in Figure 5.9. By the time $t = 1$ (20 time steps) the solution has converged to within $\sim$3% of the exact solution, $\phi = (x^2 - 1)(y^2 - 1)$.

Taking $\Delta t = 1$, the solution converges to within $\sim$1% of the exact steady state solution in only four time steps. This solution is no longer *time accurate*, but if we are concerned only with the steady state solution, approximate factorization offers a very quick means of getting to it.

5.9.1 Stability of the Factored Scheme

We will now show that the factored version of the implicit scheme is also unconditionally stable. Thus, at least for the heat equation, factorization does neither affect the order of accuracy nor the stability of the scheme. Both the von Neumann or the modified wavenumber analysis would work. With the wavenumber analysis, one assumes a solution of the form,

$$\phi_{lj}^n = \psi^n e^{ik_1 x_l} e^{ik_2 y_j}$$

for (5.45). The spatial derivative operators in (5.45) are replaced by the corresponding modified wavenumbers, $-k_1'^2, k_2'^2$ given by equation (5.34),

$$\left(1+\frac{\alpha\Delta t}{2}k_1'^2\right)\left(1+\frac{\alpha\Delta t}{2}k_2'^2\right)\psi^{(n+1)} = \left(1-\frac{\alpha\Delta t}{2}k_1'^2\right)\left(1-\frac{\alpha\Delta t}{2}k_2'^2\right)\psi^{(n)}.$$

Thus, the amplification factor is

$$\left|\frac{\psi^{(n+1)}}{\psi^{(n)}}\right| = \left|\frac{\left(1-\frac{\alpha\Delta t}{2}k_1'^2\right)\left(1-\frac{\alpha\Delta t}{2}k_2'^2\right)}{\left(1+\frac{\alpha\Delta t}{2}k_1'^2\right)\left(1+\frac{\alpha\Delta t}{2}k_2'^2\right)}\right| \le 1$$

which is always less than or equal to 1, implying that the method is unconditionally stable.

5.9.2 Alternating Direction Implicit Methods

The original split type method was introduced by Peaceman and Rachford in 1955*. Their method for an implicit solution of the 2D heat equation is of the operator splitting form rather than the factored form introduced earlier in this section. For reasons that will become apparent shortly, their method is called the alternating direction implicit (ADI) method. We will show that the ADI scheme is an equivalent formulation of the factored scheme. The following derivation of the ADI scheme is within the general scope of fractional step methods, where different terms in a partial differential equation are advanced with different time advancement schemes.

Consider the two-dimensional heat equation (5.31):

$$\frac{\partial \phi}{\partial t} = \alpha\left(\frac{\partial^2\phi}{\partial x^2}+\frac{\partial^2\phi}{\partial y^2}\right). \tag{5.50}$$

The ADI scheme for advancing this equation from step t_n to $t_n+\Delta t$ begins with splitting it into two parts: first, the equation is advanced by half the time step by a "mixed" scheme consisting of the backward Euler scheme for the $\partial^2\phi/\partial x^2$ term and explicit Euler for $\partial^2\phi/\partial y^2$; next, starting from the newly obtained solution at $t_{n+1/2}$ the roles are reversed and the backward Euler is used for the y derivative term and the explicit Euler for the x derivative term:

$$\phi^{n+1/2}-\phi^n = \frac{\alpha\Delta t}{2}\left(\frac{\partial^2\phi^{n+1/2}}{\partial x^2}+\frac{\partial^2\phi^n}{\partial y^2}\right) \tag{5.51}$$

$$\phi^{n+1}-\phi^{n+1/2} = \frac{\alpha\Delta t}{2}\left(\frac{\partial^2\phi^{n+1/2}}{\partial x^2}+\frac{\partial^2\phi^{n+1}}{\partial y^2}\right). \tag{5.52}$$

The advantage of this procedure is that at each sub-step, one has a one-dimensional implicit scheme that involves a simple tridiagonal solution as

* Peaceman, D. W., and Rachford, H. H., Jr. *SIAM J.*, **3**, 28, 1955.

opposed to the large block-tridiagonal scheme in (5.43). Note that the method is not symmetric with respect to x and y. In practice, to avoid the preferential accumulation of round-off errors in any given direction, the ordering of implicit and explicit treatments of the x and y derivatives are reversed at each time step. For example, if equations (5.51) and (5.52) are used to advance from time step n to $n+1$, then to advance from $n+1$ to $n+3/2$, backward Euler is used to advance the y derivative term and explicit Euler for the x derivative term; and then from $n+3/2$ to $n+2$, explicit Euler is used for the y derivative and backward Euler for the x derivative terms.

It is easy to show that the ADI scheme is equivalent to the factored scheme in (5.45). To do this we will first write the equations (5.51) and (5.52) using the operator notation introduced earlier:

$$\left(I - \frac{\alpha\Delta t}{2}A_x\right)\phi^{n+1/2} = \left(I + \frac{\alpha\Delta t}{2}A_y\right)\phi^n \tag{5.53}$$

$$\left(I - \frac{\alpha\Delta t}{2}A_y\right)\phi^{n+1} = \left(I + \frac{\alpha\Delta t}{2}A_x\right)\phi^{n+1/2}. \tag{5.54}$$

Equation (5.53) can be solved for $\phi^{n+1/2}$,

$$\phi^{n+1/2} = \left(I - \frac{\alpha\Delta t}{2}A_x\right)^{-1}\left(I + \frac{\alpha\Delta t}{2}A_y\right)\phi^n,$$

which is then substituted in (5.54) to yield

$$\left(I - \frac{\alpha\Delta t}{2}A_y\right)\phi^{n+1} = \left(I + \frac{\alpha\Delta t}{2}A_x\right)\left(I - \frac{\alpha\Delta t}{2}A_x\right)^{-1}\left(I + \frac{\alpha\Delta t}{2}A_y\right)\phi^n.$$

Since the $(I + \alpha\Delta t/2\,A_x)$ and $(I - \alpha\Delta t/2\,A_x)$ operators commute, we will recover (5.45):

$$\left(I - \frac{\alpha\Delta t}{2}A_x\right)\left(I - \frac{\alpha\Delta t}{2}A_y\right)\phi^{(n+1)} = \left(I + \frac{\alpha\Delta t}{2}A_x\right)\left(I + \frac{\alpha\Delta t}{2}A_y\right)\phi^{(n)}.$$

Finally, we have to address the implementation of boundary conditions. In (5.53) boundary conditions are required for $\phi^{n+1/2}$ at the two x boundaries. We refer to these boundary conditions by ϕ_B, where B can be either boundary. Peaceman and Rachford suggested using the prescribed boundary conditions for ϕ at $t = t_{n+1/2}$. Another boundary condition that is more consistent with the splitting algorithm is derived as follows.

Equations (5.53) and (5.54) are rewritten as

$$\phi^{(n+1/2)} - \frac{\alpha\Delta t}{2}A_x\phi^{(n+1/2)} = \left(I + \frac{\alpha\Delta t}{2}A_y\right)\phi^{(n)}$$

and

$$\phi^{(n+1/2)} + \frac{\alpha \Delta t}{2} A_x \phi^{(n+1/2)} = \left(I - \frac{\alpha \Delta t}{2} A_y\right) \phi^{(n+1)}.$$

Adding these two equations and evaluating at the boundaries, we obtain

$$\phi_B^{n+1/2} = 1/2 \left[\left(I + \frac{\alpha \Delta t}{2} A_y\right) \phi_B^{(n)} + \left(I - \frac{\alpha \Delta t}{2} A_y\right) \phi_B^{(n+1)}\right].$$

If there are no variations in the boundary conditions along the y direction, then the boundary condition at the intermediate step is the arithmetic mean of the boundary values at time steps n and $n+1$, which is a second-order approximation to the exact condition, $\phi(x_B, y, t_{n+1/2})$.

5.9.3 Mixed and Fractional Step Methods

Using different time advancement schemes to advance different terms in a partial differential equation has been a very powerful tool in numerical solution of complex differential equations. In the case of ADI we used this approach to avoid large matrices arising from implicit time advancement of multi-dimensional equations. This approach has also been very effective in the numerical solution of differential equations where different terms may have different characteristics (such as linear and non-linear) or different time scales. In such cases, it is most cost effective to advance the different terms using different methods.

For example, consider the Burgers equation

$$\frac{\partial u}{\partial t} + u \frac{\partial u}{\partial x} = \nu \frac{\partial^2 u}{\partial x^2}. \tag{5.55}$$

This equation has a non-linear convection-like term and a linear diffusion term. Based on our experiences with simple linear convection and diffusion equations, we know that some numerical methods are suitable for one term and not for the other. For example, the leapfrog method would probably be a good scheme for a term that has convection behavior and would not be a good choice for the diffusion phenomenon. Therefore, if we choose to advance the entire equation with leapfrog, we would probably encounter numerical instabilities. Numerical experiments have shown that this is indeed the case. Thus, it would be better to advance just the convection term with leapfrog and use another scheme for the diffusion term.

In another example, the value of ν may be such that the stability criterion for the diffusive part of the equation as given in (5.13) would impose a particularly severe restriction on the time step, which would call for an implicit scheme. But, we may not want to deal with non-linear algebraic equations, and therefore we would not want to apply it to the convection term. Let's consider explicit time advance for the convection term and an implicit scheme for the diffusion term. In fact a popular scheme for the Burgers equation is a combination of time advancement with the Adams–Bashforth method (Chapter 4), which is an explicit scheme, and the trapezoidal method for the diffusion term. This scheme is written as follows:

$$u^{n+1} - u^n = -\frac{\Delta t}{2}\left(3u^n \frac{\partial u^n}{\partial x} - u^{n-1}\frac{\partial u^{n-1}}{\partial x}\right) + \frac{\nu \Delta t}{2}\left(\frac{\partial^2 u^{n+1}}{\partial x^2} + \frac{\partial^2 u^n}{\partial x^2}\right),$$

which can be rearranged as

$$\frac{\nu}{2}\frac{\partial^2 u^{n+1}}{\partial x^2} - \frac{u^{n+1}}{\Delta t} = -\frac{u^n}{\Delta t} + \frac{1}{2}\left(3u^n \frac{\partial u^n}{\partial x} - u^{n-1}\frac{\partial u^{n-1}}{\partial x}\right) - \frac{\nu}{2}\frac{\partial^2 u^n}{\partial x^2}.$$

This is a second-order algorithm in time. Now, we can use a suitable differencing scheme for the spatial derivatives and then must solve a banded matrix at each time step. Because of explicit treatment of the non-linear terms, they appear only on the right-hand side and hence cause no difficulty.

Finally, for an interesting illustration of fractional step methods, we will consider an example of the so-called locally one dimensional (LOD) schemes. The motivation for using such schemes is the same as the approximate factorization or ADI, that is, to reduce a complex problem to a sequence of simpler ones at each time step. For example, the two-dimensional heat equation (5.31) is written as the following pair of equations:

$$\frac{1}{2}\frac{\partial u}{\partial t} = \alpha \frac{\partial^2 u}{\partial x^2} \tag{5.56}$$

$$\frac{1}{2}\frac{\partial u}{\partial t} = \alpha \frac{\partial^2 u}{\partial y^2}. \tag{5.57}$$

In advancing the heat equation from step t_n to step t_{n+1}, equation (5.56) is advanced from t_n to $t_{n+1/2}$, and (5.57) from $t_{n+1/2}$ to t_{n+1}. If the Crank–Nicolson scheme is used to advance each of the equations (5.56) and (5.57) by $\delta t/2$, then it is easy to show that this LOD scheme is identical to the ADI scheme of Peaceman and Rachford given by equations (5.53) and (5.54); the LOD scheme is just another formalism and a way of thinking about the fractional or split schemes.

5.10 Elliptic Partial Differential Equations

Elliptic equations usually arise from steady state or equilibrium physical problems. From the mathematical point of view, elliptic equations are boundary value problems where the solution is inter-related at all the points in the domain. That is, if a perturbation is introduced at one point, the solution is affected instantly in the entire domain. In other words information propagates at infinite speed in the domain of an elliptic problem. Elliptic problems are formulated in closed domains, and boundary conditions are specified on the boundary.

Standard elliptic equations include the Laplace equation,

$$\nabla^2\phi = 0, \tag{5.58}$$

the Poisson equation,

$$\nabla^2\phi = f, \tag{5.59}$$

and the Helmholtz equation,

$$\nabla^2\phi + \alpha^2\phi = 0. \tag{5.60}$$

Boundary conditions can be Dirichlet, where ϕ is prescribed on the boundary; Neumann, where the normal derivative of ϕ is prescribed on the boundary; or mixed where a combination of the two is prescribed, e.g.,

$$c_1\phi + c_2\frac{\partial\phi}{\partial n} = g, \tag{5.61}$$

where n indicates the coordinate normal to the boundary.

The numerical treatment of problems (5.58)–(5.60) are essentially identical, and for the subsequent discussion we will consider the Poisson equation in two-dimensional Cartesian coordinates. Without loss of generality, the problem is discretized in a rectangular domain in the (x, y) plane using a uniformly spaced mesh. Suppose there are $M + 1$ grid points in the x direction (x_i, $i = 0, 1, 2, 3, \ldots, M$), with $M - 1$ interior points, and the boundaries are located at x_0 and x_M respectively. Similarly, $N + 1$ points are used in the y direction. The second derivatives in the ∇^2 are approximated by second-order finite difference operators. For simplicity we will assume that $\Delta_x = \Delta_y = \Delta$. The equations for $\phi_{i,j}$ become

$$\phi_{i+1,j} - 4\phi_{i,j} + \phi_{i-1,j} + \phi_{i,j+1} + \phi_{i,j-1} = \Delta^2 f_{i,j}, \tag{5.62}$$

for $i = 1, 2, \ldots, M - 1$ and $j = 1, 2, \ldots, N - 1$.

Special treatment is required for points adjacent to the boundaries to incorporate the boundary conditions. For example, for $i = 1$ and for any $j = 2, 3, \ldots, N - 1$, equation (5.62) becomes

$$\phi_{2,j} - 4\phi_{1,j} + \phi_{1,j+1} + \phi_{1,j-1} = \Delta^2 f_{1,j} - \phi_{0,j}, \tag{5.63}$$

where we assume that $\phi_{0,j}$ is prescribed through Dirichlet boundary conditions and hence it is moved to the right-hand side. Thus, non-zero Dirichlet boundary conditions simply modify the right-hand side of (5.62). If the unknown $\phi_{i,j}$ is ordered with first increasing i, that is,

$$[\phi_{1,1}, \phi_{2,1}, \phi_{3,1}, \ldots, \phi_{M-1,1}, \phi_{1,2}, \phi_{2,2}, \phi_{3,2}, \ldots]^T,$$

then the system of equations can be written in the form

$$A\boldsymbol{x} = \boldsymbol{b}, \tag{5.64}$$

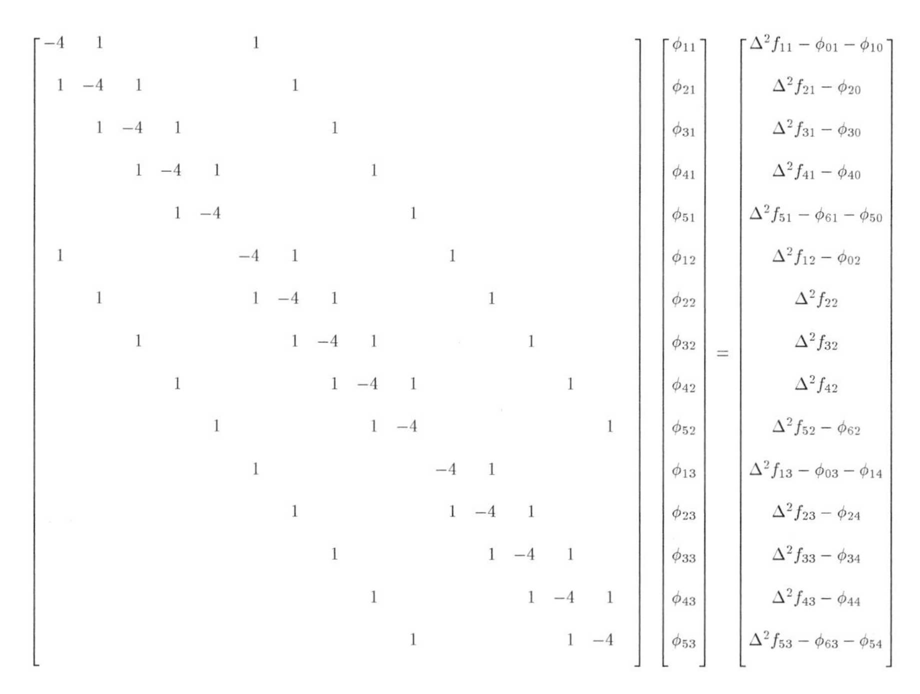

Figure 5.10 System of linear equations arising from discretizing (5.62) with $M = 6$, $N = 4$.

which is displayed in Figure 5.10 for the special case of $(M = 6,\ N = 4)$ and Dirichlet boundary conditions. The matrix A is a *block*-tridiagonal matrix similar to the one obtained in Section 5.8. The blocks are $(M - 1) \times (M - 1)$ matrices, and there are $(N - 1)$ of them on the main diagonal. Discretization with higher order schemes would lead to other block banded matrices, such as the block pentadiagonal obtained with the fourth-order central differencing.

If Neumann or mixed boundary conditions were used, then some of the matrix elements in Figure 5.10 in addition to the right-hand-side vector would have to be modified. To illustrate how this change in the system comes about, suppose that the boundary condition at $x = 0$ is prescribed to be $\partial\phi/\partial x = g(y)$, and suppose we use a second-order one-sided difference scheme to approximate this condition:

$$\frac{-3\phi_{0,j} + 4\phi_{1,j} - \phi_{2,j}}{2\Delta} = g_j.$$

By solving for $\phi_{0,j}$ using this expression, substituting in (5.63), and rearranging, we obtain

$$\frac{2}{3}\phi_{2,j} - \frac{8}{3}\phi_{1,j} + \phi_{1,j+1} + \phi_{1,j-1} = \Delta^2 f_{1,j} - \frac{2}{3}\Delta g_j.$$

It can be seen that the coefficients of $\phi_{2,j}$ and $\phi_{1,j}$ and therefore the corresponding elements of matrix A have changed in addition to the right-hand–side vector.

For this particular case of the Poisson equation in two-dimensions and with uniform mesh, the diagonal blocks are tridiagonal matrices and the sub- and super-diagonal blocks are diagonal with constant elements throughout. This property has been used to deduce efficient direct methods of solution. (A class of these methods based on Fourier expansions will be introduced in Chapter 6.) Such methods are not readily applicable for general elliptic problems in complex geometry (as opposed to, say, rectangular) with non-uniform meshes. Moreover, the matrix A is often too large for direct inversion techniques. Alternatives to direct methods are the highly popular iterative methods, which we will discuss next.

5.10.1 Iterative Solution Methods

In this and the subsequent sections, we consider the solution of equation (5.64) by iterative techniques. In fact the methodology that will be developed is for solving general systems of linear algebraic equations, $A\boldsymbol{x} = \boldsymbol{b}$, which may or may not have been derived from a particular partial differential equation. In solving a system of algebraic equations iteratively, one begins with a "guess" for the solution, and uses an algorithm to iterate on this guess which hopefully improves the solution. In contrast to Gauss elimination where the exact solution of a system of linear equations is obtained (to within computer round-off error), with iterative methods an approximate solution to a prescribed accuracy is sought. In the problems of interest in this chapter, where the system of algebraic equations is obtained from numerical approximation (discretization) of a differential equation, the choice of iterative methods over Gauss elimination is further justified by realizing that the equations represent an approximation to the differential equation and therefore it would not be necessary to obtain the exact solution of approximate equations. The expectation is that accuracy improves by increasing the number of iterations; that is, the method *converges* to the exact solution as the number of iterations increases. Moreover, matrices obtained from discretizing PDEs are usually sparse (a lot more zero than non-zero elements) and iterative methods are particularly advantageous in memory requirements with such systems.

Consider (5.64), and let $A = A_1 - A_2$. Equation (5.64) can be written as

$$A_1\boldsymbol{x} = A_2\boldsymbol{x} + \boldsymbol{b}. \tag{5.65}$$

An iterative solution technique is constructed as follows:

$$A_1\boldsymbol{x}^{(k+1)} = A_2\boldsymbol{x}^{(k)} + \boldsymbol{b}, \tag{5.66}$$

where $k = 0, 1, 2, 3, \ldots$ is the iteration index. Starting from an initial guess for the solution $\boldsymbol{x}^{(0)}$, equation (5.66) is used to solve for $\boldsymbol{x}^{(1)}$, which is then used to find $\boldsymbol{x}^{(2)}$, and so on. For the algorithm (5.66) to be viable, the following requirements must be imposed:

1. A_1 should be easily "invertible." Otherwise, at each iteration we are faced with solving a system of equations that can be as difficult as the original system, $A\boldsymbol{x} = \boldsymbol{b}$.
2. Iterations should converge (hopefully rapidly), that is,

$$\lim_{k\to\infty} \boldsymbol{x}^{(k)} = \boldsymbol{x}.$$

We will first establish a criterion for convergence. Let the error at the kth iteration be denoted by $\boldsymbol{\epsilon}^{(k)}$:

$$\boldsymbol{\epsilon}^{(k)} = \boldsymbol{x} - \boldsymbol{x}^{(k)}.$$

Subtracting (5.65) from (5.66) leads to

$$A_1\boldsymbol{\epsilon}^{(k+1)} = A_2\boldsymbol{\epsilon}^{(k)}$$

or

$$\boldsymbol{\epsilon}^{(k+1)} = A_1^{-1}A_2\boldsymbol{\epsilon}^{(k)}.$$

From this expression we can easily deduce that the error at iteration k is related to the initial error via

$$\boldsymbol{\epsilon}^{(k)} = \left(A_1^{-1}A_2\right)^k\boldsymbol{\epsilon}^{(0)}. \tag{5.67}$$

For convergence we must have

$$\lim_{k\to\infty} \boldsymbol{\epsilon}^{(k)} = 0.$$

We know from linear algebra (see Appendix) that this will happen if

$$\rho = |\lambda_i|_{\max} \leq 1, \tag{5.68}$$

where λ_i are the eigenvalues of the matrix $A_1^{-1}A_2$. ρ is called the *spectral radius* of convergence of the iterative scheme and is related to its rate of convergence. The performance of any iterative scheme and its *rate* of convergence are directly connected to the matrix A and its decomposition into A_1 and A_2.

5.10.2 The Point Jacobi Method

The simplest choice for A_1 is the diagonal matrix D consisting of the diagonal elements of A, a_{ii}. Surely, a diagonal matrix satisfies the first requirement that it be easily invertible. For the matrix of Figure 5.10, A_1 would be the diagonal matrix with -4 on the diagonal. A_1^{-1} is readily computed to be the diagonal matrix with $-1/4$ on the diagonal. A_2 can be deduced from th matrix of Figure 5.10 by replacing every 1 with -1 and each -4 with zero. Thus, application of the point Jacobi method to the system of equations in Figure 5.10 leads to the following iterative scheme:

$$\phi^{(k+1)} = -\frac{1}{4} A_2 \phi^{(k)} - \frac{1}{4} \boldsymbol{R}, \tag{5.69}$$

where $\boldsymbol{R}$ is the right-hand vector in Figure 5.10. Using the index notation, equation (5.69) can be written as follows:

$$\phi_{ij}^{(k+1)} = \frac{1}{4}\left[\phi_{i-1,j}^{(k)} + \phi_{i+1,j}^{(k)} + \phi_{i,j-1}^{(k)} + \phi_{i,j+1}^{(k)}\right] - \frac{1}{4} R_{ij}, \tag{5.70}$$

where the indices i and j are used in the same order as in the ϕ column of Figure 5.10. Starting with an initial guess $\phi_{ij}^{(0)}$, subsequent approximations, $\phi_{ij}^{(1)}, \phi_{ij}^{(2)}, \ldots,$ are easily computed from (5.70). Note that application of the point Jacobi does not involve storage or manipulation with any matrices. One simply updates the value of ϕ at the grid point (ij) using a simple average of the surrounding values (north, south, east, and west) from the previous iteration.

For convergence, the eigenvalues of the matrix $A_1^{-1} A_2 = -1/4\, A_2$ must be computed. For this particular example, it can be shown using a discrete analog of the method of separation of variables (used to solve partial differential equations analytically) that the eigenvalues are

$$\lambda_{mn} = \frac{1}{2}\left[\cos\frac{m\pi}{M} + \cos\frac{n\pi}{N}\right] \quad m = 1, 2, 3, \ldots, M-1$$
$$n = 1, 2, 3, \ldots, N-1. \tag{5.71}$$

It is clear that $|\lambda_{mn}| < 1$ for all m and n, and the method converges. The eigenvalue with the largest magnitude determines the rate of convergence*. For large M and N, we expand the cosines in equation (5.71) (with $n = m = 1$) in power series, and to leading order we get

$$|\lambda|_{\max} = 1 - \frac{1}{4}\left[\frac{\pi^2}{M^2} + \frac{\pi^2}{N^2}\right] + \cdots$$

Thus, for large M and N, $|\lambda|_{\max}$ is only slightly less than 1, and the convergence is very slow. This is why the point Jacobi method is rarely used in practice, but it does provide a good basis for development and comparison with improved methods.

* This can be seen by eigenvectors diagonalization of the matrix $A_1^{-1} A_2$. For defective systems (matrices without a complete set of eigenvectors), unitary triangularization can be used to prove the same result. The reader is referred to the Appendix and standard textbooks in linear algebra for these matrix transformations.

EXAMPLE 5.7 Number of Iterations for Specified Accuracy

How many iterations are required to reduce the initial error in the solution of a Poisson equation by a factor of 10^{-m} using the point Jacobi method? Let n be the required number of iterations and $B = A_1^{-1}A_2$ in (5.67). Taking the norm of both sides of (5.67) and using the norm properties (see Appendix), we obtain

$$\begin{aligned}\left\|\epsilon^{(n)}\right\| &= \left\|B^n\epsilon^{(0)}\right\| \\ &\le \left\|B^n\right\|_2 \left\|\epsilon^{(0)}\right\| \\ &\le \left\|B\right\|_2^n \left\|\epsilon^{(0)}\right\|.\end{aligned}$$

Since B is symmetric, it can be shown that $\|B\|_2 = |\lambda|_{\max}$. Thus

$$\left\|\epsilon^{(n)}\right\| \le |\lambda|_{\max}^n \left\|\epsilon^{(0)}\right\|.$$

To reduce the error by factor of 10^{-m}, we should have

$$|\lambda|_{\max}^n \le 10^{-m}.$$

Taking the logarithms of both sides and solving for n

$$n \ge \frac{-m}{\log|\lambda|_{\max}},$$

where we have taken into account that $\log|\lambda_i| < 0$ by reversing the direction of the inequality. For example, suppose in a rectangular domain we use $M = 20$ and $N = 20$, then

$$\lambda_{\max} = \cos\frac{\pi}{20} = 0.988.$$

To reduce the initial error by a factor of 1000, i.e., $m = 3$, we require 558 iterations. For $M = N = 100$, about 14000 iterations would be required to reduce the error by a factor of 1000.

In the next two sections we will discuss methods that improve on the point Jacobi scheme.

5.10.3 Gauss–Seidel Method

Consider the point Jacobi method in equation (5.70), which is a recipe for computation of $\phi_{i,j}^{(k+1)}$ given all the data at iteration k. Implementation of (5.70) in a computer program consists of a loop over k and two inner loops over indices i and j. Clearly, $\phi_{i-1,j}^{(k+1)}$ and $\phi_{i,j-1}^{(k+1)}$ are computed before $\phi_{i,j}^{(k+1)}$. Thus, in equation (5.70) instead of using $\phi_{i-1,j}^{(k)}$ and $\phi_{i,j-1}^{(k)}$, one can use their *updated* values, which are presumably more accurate. This gives us the formula for the Gauss–Seidel method:

$$\phi_{ij}^{(k+1)} = \frac{1}{4}\left[\phi_{i-1,j}^{(k+1)} + \phi_{i+1,j}^{(k)} + \phi_{i,j-1}^{(k+1)} + \phi_{i,j+1}^{(k)}\right] - \frac{1}{4}R_{ij}. \tag{5.72}$$

In the matrix splitting notation of Section 5.10.1,

$$A = A_1 - A_2,$$

where for Gauss–Seidel

$$A_1 = D - L \quad \text{and} \quad A_2 = U, \tag{5.73}$$

D is the diagonal matrix consisting of the diagonal elements of A, L is the lower triangular matrix consisting of the negative of the lower triangular elements of A, and U is an upper triangular matrix consisting of the negative of the upper triangular elements of A. The matrices L and U are not to be confused with the usual LU decomposition of A discussed in the context of Gauss elimination in linear algebra (see Appendix). Since A_1 is lower triangular, the requirement (1) in Section 5.10.1 is met (even though more operations are required to invert a lower triangular matrix than a diagonal one). It turns out that for the discrete Poisson equation considered in Section 5.10, the eigenvalues of the matrix $A_1^{-1}A_2$ are simply squares of the eigenvalues of the point Jacobi method, i.e.,

$$\lambda_{mn} = \frac{1}{4}\left[\cos\frac{m\pi}{M} + \cos\frac{n\pi}{N}\right]^2 \quad m = 1, 2, 3, \ldots, M-1$$
$$n = 1, 2, 3, \ldots, N-1. \tag{5.74}$$

Thus, the Gauss–Seidel method converges twice as fast as the point Jacobi method (see Example 5.7) and hence would require *half* as many iterations as the point Jacobi method to converge to within a certain error tolerance.

5.10.4 Successive Over Relaxation Scheme

One of the most successful iterative methods for the solution of a system of algebraic equations is the successive over relaxation (SOR) method. This method attempts to increase the rate of convergence of the Gauss–Seidel method by introducing a parameter into the iteration scheme and then optimizing it for fast convergence. We have already established that the rate of convergence depends on the largest eigenvalue of the iteration matrix, $A_1^{-1}A_2$. Our objective is then to find the optimal parameter to reduce as much as possible the largest eigenvalue. Consider the Gauss–Seidel method for the solution of (5.66) with A_1 and A_2 given by (5.73):

$$(D - L)\phi^{(k+1)} = U\phi^{(k)} + \boldsymbol{b}. \tag{5.75}$$

Let the change in the solution between two successive iterations be denoted by

$$\boldsymbol{d} = \phi^{(k+1)} - \phi^{(k)}.$$

Thus, for Gauss–Seidel, or for that matter, any iterative method, we have the following identity:

$$\phi^{(k+1)} = \phi^{(k)} + \boldsymbol{d}.$$

We now attempt to increase (accelerate) the change between two successive iterations by using an acceleration parameter; that is,

$$\phi^{(k+1)} = \phi^{(k)} + \omega \boldsymbol{d}, \tag{5.76}$$

where $\omega > 1$ is the acceleration or "relaxation" parameter. Note that if ω were less than 1 we would be decelerating (reducing) the change at each iteration; with $\omega = 1$ the Gauss–Seidel method is recovered. Thus, in SOR one first uses the Gauss–Seidel method (5.75) to compute an intermediate solution, $\tilde{\phi}$:

$$D\tilde{\phi}^{(k+1)} = L\phi^{(k+1)} + U\phi^{(k)} + \boldsymbol{b}. \tag{5.77}$$

We do yet not accept this as the solution at the next iteration; we want to increase the incremental change from the previous iteration. The SOR solution at the next iteration is then given by

$$\phi^{(k+1)} = \phi^{(k)} + \omega\left(\tilde{\phi}^{(k+1)} - \phi^{(k)}\right), \tag{5.78}$$

where the relaxation parameter ω is yet to be determined and hopefully optimized. To study the convergence properties of the method, we eliminate $\tilde{\phi}^{(k+1)}$ between equations (5.77) and (5.78) and solve for $\phi^{(k+1)}$:

$$\phi^{(k+1)} = \underbrace{(I - \omega D^{-1}L)^{-1}[(1-\omega)I + \omega D^{-1}U]}_{G_{\text{SOR}}}\phi^{(k)} + (I - \omega D^{-1}L)^{-1}\omega D^{-1}\boldsymbol{b}.$$

Convergence is dependent on the eigenvalues of the matrix G_{SOR} which is the iteration matrix, $A_1^{-1}A_2$, for SOR. It can be shown that for the discretized Poisson operator, the eigenvalues are given by

$$\lambda^{\frac{1}{2}} = \frac{1}{2}\left(\pm|\mu|\omega \pm \sqrt{\mu^2\omega^2 - 4(\omega - 1)}\right), \tag{5.79}$$

where μ is an eigenvalue of the point Jacobi matrix, $G_J = D^{-1}(L + U)$.

To optimize convergence, one should select the relaxation parameter ω to minimize the largest eigenvalue λ (we choose plus signs in (5.79)). It turns out that $d\lambda/d\omega = 0$ does not have a solution, but the corresponding functional relationship (5.79) has an absolute minimum when $d\lambda/d\omega$ is infinite (see Figure 5.11). At this point, the argument under the square root in (5.79) is zero.

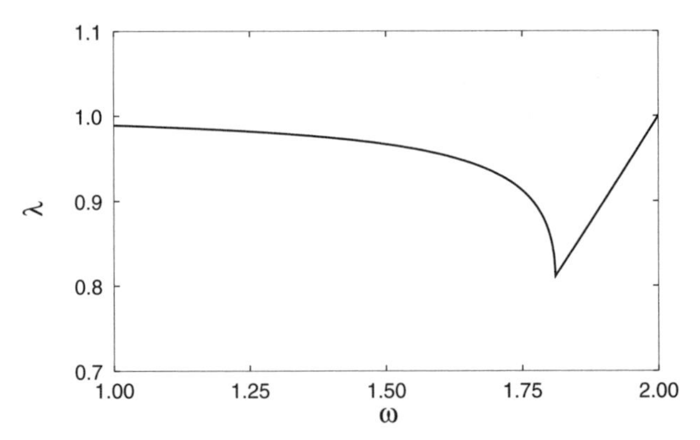

Figure 5.11 The eigenvalues λ of the matrix G_{SOR} plotted versus ω according to (5.79) with $\mu_{max} = 0.9945$. This value of μ_{max} corresponds to a 31×31 mesh and is obtained from (5.71) using $M = N = 30$ and $m = n = 1$.

Thus, the minimum of the largest eigenvalue occurs at

$$\omega_{opt} = \frac{2}{1 + \sqrt{1 - \mu_{max}^2}} \tag{5.80}$$

where μ_{max} is the largest eigenvalue of the Point–Jacobi method. Recall that $|\mu_{max}|$ is just slightly less than 1 and therefore ω_{opt} is just under 2. The optimum value of ω usually used is between 1.7 and 1.9. The precise value depends on μ_{max} and therefore on the number of grid points used. For problems with irregular geometry and non-uniform mesh, ω_{opt} cannot be obtained analytically but must be found by numerical experiments. For example, to solve a Poisson equation several times with different right-hand sides, first obtain ω by numerical experiments and then use it for the "production runs."

EXAMPLE 5.8 Iterative Solution of an Elliptic Equation

We again consider the problem of Example 5.6, but now we will solve it by iteration rather than time advancing the solution to steady state. The steady state PDE is the Poisson equation

$$-\nabla^2\phi = q \qquad q = 2(2 - x^2 - y^2)$$

with the boundary conditions

$$\phi(\pm 1, y) = 0 \qquad \phi(x, \pm 1) = 0.$$

No initial condition is required as the problem is no longer time dependent. We will choose as an initial guess for our iterative solution $\phi^{(0)}(x, y) = 0$. The problem will be solved with the point Jacobi, Gauss–Seidel, and SOR algorithms. Spatial derivatives are calculated with second-order central differences ($\Delta x = \Delta y = \Delta$).

$$\frac{\phi_{i+1,j} - 2\phi_{i,j} + \phi_{i-1,j}}{\Delta^2} + \frac{\phi_{i,j+1} - 2\phi_{i,j} + \phi_{i,j-1}}{\Delta^2} = -q_{i,j}.$$

With k specifying the iteration level, the different algorithms are

1. Point Jacobi

$$\phi_{i,j}^{(k+1)} = \frac{1}{4}\left[\phi_{i+1,j}^{(k)} + \phi_{i-1,j}^{(k)} + \phi_{i,j+1}^{(k)} + \phi_{i,j-1}^{(k)}\right] + \frac{\Delta^2}{4} q_{i,j}.$$

2. Gauss–Seidel

$$\phi_{i,j}^{(k+1)} = \frac{1}{4}\left[\phi_{i+1,j}^{(k)} + \phi_{i-1,j}^{(k+1)} + \phi_{i,j+1}^{(k)} + \phi_{i,j-1}^{(k+1)}\right] + \frac{\Delta^2}{4} q_{i,j}.$$

3. Successive over relaxation

$$\tilde{\phi}_{i,j} = \frac{1}{4}\left[\phi_{i+1,j}^{(k)} + \tilde{\phi}_{i-1,j} + \phi_{i,j+1}^{(k)} + \tilde{\phi}_{i,j-1}\right] + \frac{\Delta^2}{4} q_{i,j}$$

$$\phi_{i,j}^{(k+1)} = \phi_{i,j}^{(k)} + \omega\left(\tilde{\phi}_{i,j} - \phi_{i,j}^{(k)}\right).$$

The number of iterations needed to bring each solution to within 0.01% of the exact solution are shown in the table:

Method	**Iterations**
Point Jacobi	675
Gauss–Seidel	339
SOR ($\omega = 1.8$)	169

The SOR method is probably the first example of a procedure where the convergence of an iterative scheme is enhanced by clever manipulation of the eigenvalues of the iteration matrix, $A_1^{-1}A_2$. A variant of this procedure, referred to as *pre-conditioning*, has received considerable attention in numerical analysis. In its simplest form, one pre-multiplies the system of equations at hand by a carefully constructed matrix that yields a more favorable eigenvalue spectrum for the iteration matrix.

5.10.5 Multigrid Acceleration

One of the most powerful acceleration schemes for the convergence of iterative methods in solving elliptic problems is the multigrid algorithm. The method is based on the realization that different components of the solution converge to the exact solution at different rates and hence should be treated differently. Suppose the residual or the error vector in the solution is represented as a linear combination of a set of basis vectors which when plotted on the grid would range from smooth to rapidly varying (just like low- and high-frequency sines and cosines). It turns out that, as the examples below will demonstrate, the smooth component of the residual converges very slowly to zero and the rough part converges quickly. The multigrid algorithm takes advantage of this to substantially reduce the overall effort required to obtain a converged solution.

Recall that our objective was to solve the equation

$$A\phi = \boldsymbol{b},$$

where A is a matrix obtained from a finite difference approximation to a differential equation. Let $\psi = \phi^{(n)}$ be an approximation to the solution ϕ, which is obtained from an iterative scheme after n iterations. The residual vector $\boldsymbol{r}$ is defined as

$$A\psi = \boldsymbol{b} - \boldsymbol{r}. \tag{5.81}$$

The residual approaches zero if the approximate solution ψ approaches the exact solution ϕ. Subtracting these two equations leads to an equation for the error $\epsilon = \phi - \psi$ in terms of the residual $\boldsymbol{r}$

$$A\epsilon = \boldsymbol{r}, \tag{5.82}$$

which is called the residual equation. Clearly, as the residual goes to zero, so does the error and vice versa. Accordingly, we often talk about driving the residual to zero in our iterative solution process, and we measure the performance of a given solution procedure in terms of the number of iterations required to drive the residual to zero.

For illustration purposes, consider the one-dimensional boundary value problem:

$$\frac{d^2u}{dx^2} = \sin k\pi x \quad 0 \le x \le 1 \tag{5.83}$$

$$u(0) = u(1) = 0.$$

The integer k is called the wavenumber and is an indicator of how many oscillations the sine wave would go through in the domain. Higher values of k correspond to more oscillations or "rougher" behavior. The exact solution is, of course, $u = -1/k^2\pi^2 \sin k\pi x$; but we will pretend we don't know this and embark on solving the problem using a finite difference approximation on $N+1$ uniformly spaced grid points of size $h = 1/N$:

$$\frac{u_{j+1} - 2u_j + u_{j-1}}{h^2} = \sin k\pi x_j \quad j = 1, 2, \ldots, N-1 \tag{5.84}$$

$$u_0 = u_N = 0.$$

Suppose, as we would do in real world non-trivial problems, we start the iterative process with a completely ignorant initial guess, $\boldsymbol{u}^{(0)} = 0$. From (5.81), the initial residual is $r_j = \sin k\pi jh$. We will use the Gauss–Seidel as the basic iteration scheme which, when applied to the original equation, takes the form

$$u_j^{(n+1)} = \frac{1}{2}\left[u_{j+1}^{(n)} + u_{j-1}^{(n+1)} - h^2 \sin k\pi jh\right],$$

where n is the iteration index.

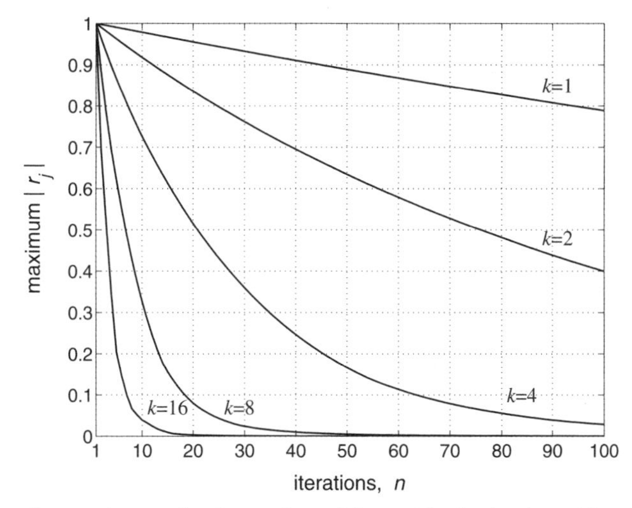

Figure 5.12 The maximum absolute value of the residual $\boldsymbol{r}$ (at the grid points) against the number of iterations for the solution of (5.84) with $N = 64$, using several values of k.

Figure 5.12 shows the evolution of the maximum residual, $\boldsymbol{r} = \boldsymbol{b} - A\boldsymbol{u}^{(n)}$, with the number of iterations for different values of wavenumber k. It is clear that the convergence is faster for higher values of k. That is, the residual, and hence the error, goes to zero faster for more rapidly varying right-hand sides. Now, consider a slightly more complicated right-hand side for (5.83):

$$\frac{d^2u}{dx^2} = \frac{1}{2}[\sin \pi x + \sin 16\pi x] \tag{5.85}$$

$$u_0 = u_N = 0.$$

The residual as a function of the number of iterations is shown in Figure 5.13. Notice that, initially, the residual goes down rapidly and then it virtually stalls. *This type of convergence history is observed frequently in practice when standard iterative schemes are used.* The reason for this behavior is that the rapidly varying part of the residual goes to zero quickly and the smooth part of it remains and as we have seen in the previous example, diminishes slowly. The initial residual, which is the same as the right-hand side of the differential equation, and its profile after 10 and 100 iterations are shown in Figure 5.14. Clearly only the smooth part of the residual has remained after 100 iterations.

Perhaps the key observation in the development of the multigrid algorithm is that a slowly varying function on a fine grid would appear as a more rapidly varying function (or rougher) on a coarse grid. This can be illustrated quantitatively by considering $\sin k\pi x$ evaluated on $N + 1$ grid points in $0 \leq x \leq 1$:

$$\sin k\pi x_j = \sin k\pi jh = \sin \frac{k\pi j}{N}.$$

Let N be even. The range of wavenumbers k that can be represented on this grid is $1 \leq k \leq N - 1$. A sine wave with wavenumber $k = N/2$ has a wavelength

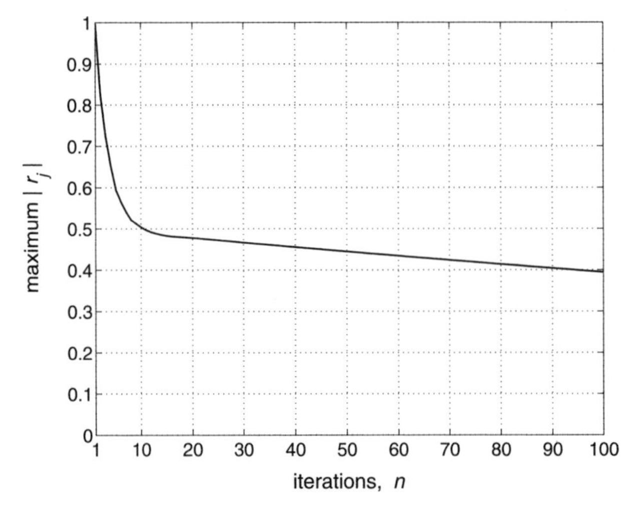

Figure 5.13 The maximum absolute value of the residual $\boldsymbol{r}$ (at the grid points) against the number of iterations for the solution of the finite difference approximation to (5.85) with $N = 64$.

equal to four grid points, where the grid points are at the maxima, minima, and the zero crossings. Let $k = k_m$ be in the first half of wavenumbers allowed, i.e., $1 \leq k_m \leq N/2$. The values of $\sin k_m \pi x_j$ evaluated at the even-numbered grid points are

$$\sin \frac{2k_m \pi j}{N} = \sin \frac{k_m \pi j}{N/2},$$

which is identical to the same function discretized on the coarse grid of $N/2 + 1$ points, but now k_m belongs to the upper half of the wavenumbers allowed on this coarse grid. Therefore, a relatively low wavenumber sine function on a fine

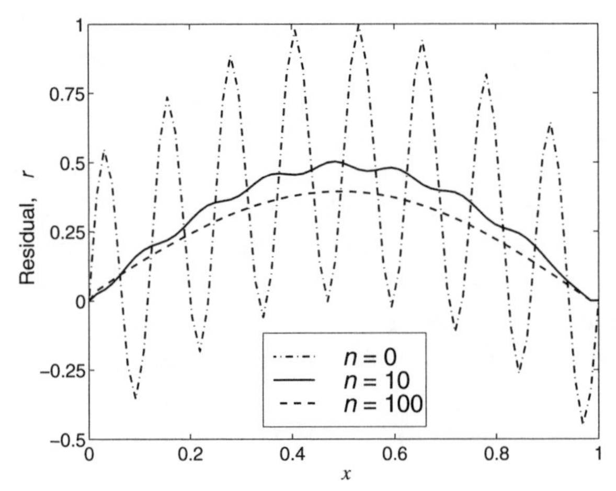

Figure 5.14 The residual at iteration numbers 0, 10, and 100 for the solution of the finite difference approximation to 5.85 with $N = 64$.

grid appears as a relatively high wavenumber sine function on a coarse grid of half the size.

Thus, according to our earlier observations of the convergence rates of iterative solutions, we might get faster convergence on the smooth part of the solution, if we transfer the problem to a coarse grid. And since the smooth part of the solution does not require many grid points to be represented, such a transfer would not incur a large error. This is the multigrid strategy: as soon as the convergence of the residual stalls (as in Figure 5.13), the iterative process is transferred to a coarse grid. On the coarse grid, the smooth part of the residual is annihilated faster and cheaper (because of fewer grid points); after this is accomplished, one can interpolate the residual back to the fine grid and work on the high wavenumber parts. This back and forth process between the fine and coarse grids continues until overall convergence is achieved. In transferring data from fine grid to coarse grid (called *restriction*) we can simply take every other data point. For transfer between coarse and fine grid (called *prolongation*) we can use a straightforward linear interpolation.

The basic dual-grid multigrid algorithm is summarized below:

1. Perform a few iterations on the original equation, $A\boldsymbol{\phi} = \boldsymbol{b}$, on the fine grid with the mesh spacing h. Let the resulting solution be denoted by $\boldsymbol{\psi}$. Calculate the residual $\boldsymbol{r} = \boldsymbol{b} - A\boldsymbol{\psi}$ on the same grid.
2. Transfer the residual to a coarse grid (restriction) of mesh spacing $2h$, and on this grid iterate on the error equation $A\boldsymbol{\epsilon} = \boldsymbol{r}$, with the initial guess $\boldsymbol{\epsilon}^0 = 0$.
3. Interpolate (prolongation) the resulting $\boldsymbol{\epsilon}$ to the fine grid. Make a correction on the previous $\boldsymbol{\psi}$ by adding it to $\boldsymbol{\epsilon}$, i.e., $\boldsymbol{\psi}_{\text{new}} = \boldsymbol{\psi} + \boldsymbol{\epsilon}$. Use $\boldsymbol{\psi}_{\text{new}}$ as the initial guess to iterate on the original problem, $A\boldsymbol{\phi} = \boldsymbol{b}$.
4. Repeat the process.

Another point that comes to mind is why stop at only one coarse grid? After a few iterations on a coarse grid where some of the low-frequency components of the residual are reduced, we can move on to yet a coarser grid, perform a few iterations there and so on. In fact the coarsest grid that can be considered is a grid of one point where we can get the solution directly and then work backward to finer and finer grids. When we return to the finest grid, and if the residual has not sufficiently diminished, we can repeat the whole process again. This recursive thinking and the use of a hierarchy of grids (each half the size of the previous one) is a key part of all multigrid codes. Three recursive approaches to multigrid are illustrated in Figure 5.15. Figure 5.15(a) shows the recursive algorithm that we just discussed and is referred to as the V cycle . The other two sketches in Figure 5.15 illustrate the so-called W cycle and the full multigrid cycle (FMC). In FMC one starts the problem on the coarsest grid and uses the result as the initial condition for the finer mesh and so on. After reaching the finest grid one usually proceeds with the W cycle.

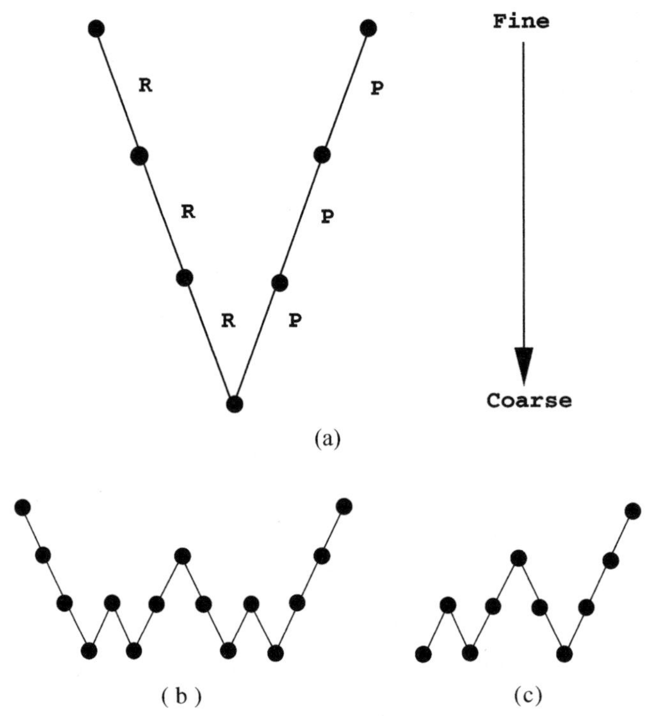

Figure 5.15 Grid selection for (a) V cycle, (b) W cycle, and (c) full multigrid cycle (FMC) algorithms. R refers to restriction or transfer from fine to coarse grid, P refers to prolongation or transfer from coarse to fine grid.

EXAMPLE 5.9 One-Dimensional V Cycle Multigrid

We now solve the boundary value problem in (5.85) using a V cycle multigrid algorithm with Gauss–Seidel as the basic iteration scheme. The finest grid has $N = N_0 = 64$, the coarsest grid has $N = 2$ (one unknown), and each of the other grids has half the value of N of the previous one. At each grid, the iteration formula is

$$u_j^{(n+1)} = \frac{1}{2}\left[u_{j+1}^{(n)} + u_{j-1}^{(n+1)}\right] - \frac{h^2}{4}[\sin \pi j h + \sin 16\pi j h] \quad j = 1, \ldots, N-1, \tag{5.86}$$

where n is the iteration index and $h = 1/N$. The initial guess is $\boldsymbol{u}^{(0)} = 0$, for $N = 64$. At each node of the V cycle, only one Gauss–Seidel iteration is performed, meaning that n takes only the value zero in the formula above. The residual $\boldsymbol{r}$ is restricted from a grid of mesh spacing h to a grid of mesh spacing $2h$ according to

$$r_j^{2h} = \frac{1}{4}\left(r_{2j-1}^h + 2r_{2j}^h + r_{2j+1}^h\right) \quad j = 1, \ldots, N/2 - 1,$$

where $N/2 + 1$ is the total number of points on the coarser grid; the superscripts indicate the grid of the corresponding mesh spacing. Working

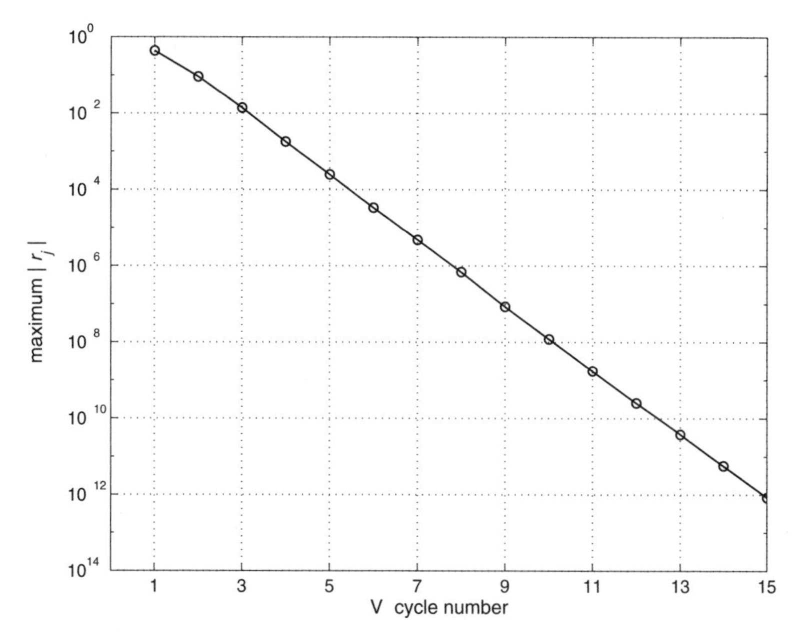

Figure 5.16 The maximum absolute value of the residual $\boldsymbol{r}$ (at the grid points) after each V cycle in Example 5.9.

backward to finer grids, the error is interpolated linearly

$$\epsilon_{2j}^{h} = \epsilon_{j}^{2h} \quad j = 0, \ldots, N$$

$$\epsilon_{2j+1}^{h} = \frac{\epsilon_{j}^{2h} + \epsilon_{j+1}^{2h}}{2} \quad j = 0, \ldots, N-1,$$

where $2N+1$ is the total number of points on the finer grid. The whole V cycle is repeated 15 times. The maximum absolute value of the residual at the end of each V cycle is plotted in Figure 5.16. The number of times the right-hand side of (5.86) is evaluated in one V cycle is

$$2[(N_0 - 1) + (N_0/2 - 1) + \cdots + (N_0/16 - 1)] + (N_0/32 - 1),$$

which is $(31/8)N_0 - 6 = 242$ for $N_0 = 64$. We see from Figure 5.16 that it takes five V cycles for the maximum value of the residual to drop below 10^{-3}. If the calculations used to obtain Figure 5.13 (Gauss–Seidel scheme without multigrid) were continued, we would need 2580 iterations for the residual to drop below 10^{-3}. This means $(2580 \times 63)/(5 \times 242) \approx 134$ times more work. The power of multigrid acceleration is evident.

Note that if the residual $\boldsymbol{r}$ is restricted by simply taking every other point from the finer grid,

$$r_{j}^{2h} = r_{2j}^{h} \quad j = 1, \ldots, N/2 - 1,$$

we would need more iterations for the residual to drop to a certain value. In the present example the residual would drop below 10^{-12} after 27 V cycles, compared to 15 V cycles in Figure 5.16.

EXAMPLE 5.10 V Cycle Multigrid for the Poisson Equation

We apply the V cycle multigrid algorithm to the Poisson equation of Example 5.8. We use the same procedure as in the previous example with the following changes. The finest grid has 33×33 total points. Three Gauss–Seidel iterations are performed at each node of the V cycle. The residual $\boldsymbol{r}$ is restricted according to

$$r_{ij}^{2h} = \frac{1}{16}\Big[r_{2i-1,2j-1}^{h} + r_{2i+1,2j-1}^{h} + r_{2i-1,2j+1}^{h} + r_{2i+1,2j+1}^{h}$$
$$+ 2\big(r_{2i,2j-1}^{h} + r_{2i,2j+1}^{h} + r_{2i-1,2j}^{h} + r_{2i+1,2j}^{h}\big) + 4r_{2i,2j}^{h}\Big]$$
$$i, j = 1, \ldots, N/2 - 1.$$

The error is interpolated according to

$$\epsilon_{2i,2j}^{h} = \epsilon_{ij}^{2h}$$
$$\epsilon_{2i+1,2j}^{h} = \frac{1}{2}\big(\epsilon_{ij}^{2h} + \epsilon_{i+1,j}^{2h}\big)$$
$$\epsilon_{2i,2j+1}^{h} = \frac{1}{2}\big(\epsilon_{ij}^{2h} + \epsilon_{i,j+1}^{2h}\big)$$
$$\epsilon_{2i+1,2j+1}^{h} = \frac{1}{4}\big(\epsilon_{ij}^{2h} + \epsilon_{i+1,j}^{2h} + \epsilon_{i,j+1}^{2h} + \epsilon_{i+1,j+1}^{2h}\big).$$

Twenty-five fine grid iterations (one initial Gauss–Seidel iteration and four V cycles) were needed to bring the solution to within 0.01% of the exact solution. In Example 5.8, the Gauss–Seidel scheme needed 339 iterations.

There is a lot more to multigrid than we can discuss in this book in terms of variations to the basic algorithm, programming details, and analysis. Fortunately, a wealth of literature exists on multigrid methods as applied to many partial differential equations that the reader can consult.

A side benefit of our discussions in this section was the preview provided of the power of a tool of analysis that one has when thinking about the various components of the algorithm and their dynamics in terms of Fourier modes. In the next chapter, we will introduce a new brand of numerical analysis based on Fourier and other modal decompositions.

EXERCISES

1. Use the modified wavenumber analysis to show that the application of the second-order upwind spatial differencing scheme

$$\left.\frac{\partial^2 \phi}{\partial x^2}\right|_j = \frac{-\phi_{j+3} + 4\phi_{j+2} - 5\phi_{j+1} + 2\phi_j}{\Delta x^2}$$

to the heat equation would lead to numerical instability.

2. Give the details of a second-order numerical scheme for the 1D heat equation in the domain $0 \leq x \leq 1$ with the following boundary conditions (encountered in problems with mixed convection and conduction heat transfer):

$$\phi = 1 \quad \text{at} \quad x = 0, \qquad \text{and} \quad a\phi + b\frac{\partial \phi}{\partial x} = c \quad \text{at} \quad x = 1.$$

Formulate the problem for both explicit and implicit time advancements. In the latter case show how the derivative boundary condition would change the matrix elements. In the text we discussed a similar problem where derivative boundary conditions were evaluated using one-sided finite differences.
Note: Another method of implementing derivative boundary conditions is by placing an "artificial" point outside the domain (in this case, just outside of $x = 1$), the equations and boundary conditions are then enforced at the physical boundary.

3. Use the von Neumann analysis to show that the Du Fort–Frankel scheme is unconditionally stable. This problem cannot be done analytically, the von Neumann analysis leads to a quadratic equation for the amplification factor. The amplification factor is a function of $\gamma = \alpha \Delta t / \Delta x^2$ and the wavenumber. Stability can be demonstrated by plotting the amplification factor for different values of γ as a function of wavenumber.

4. Suppose the 1D convection equation (5.11) is advanced in time by the leapfrog method and for spatial differencing either the second-order central differencing or the fourth-order Padé scheme is used. Compare the maximum CFL numbers for the two spatial differencing schemes. How does CFL_{max} change with increasing spatial accuracy?

5. The following numerical method has been proposed to solve $\frac{\partial u}{\partial t} = c\frac{\partial u}{\partial x}$:

$$\frac{1}{\Delta t}\left[u_j^{n+1} - \frac{1}{2}\left(u_{j+1}^n + u_{j-1}^n\right)\right] = \frac{c}{2\Delta x}\left[u_{j+1}^n - u_{j-1}^n\right].$$

(a) Find the range of CFL number $c\Delta t/\Delta x$ for which the method is stable.
(b) Is the method consistent (i.e., does it reduce to the original PDE as Δx, $\Delta t \to 0$)?

6. The Douglas Rachford ADI scheme for the 3D heat equation is given by

$$\left(I - \frac{\alpha \Delta t}{2} A_x\right)\phi^* = \left[I + \frac{\alpha \Delta t}{2}(A_y + A_z)\right]\phi^n$$

$$\left(I - \frac{\alpha \Delta t}{2} A_y\right)\phi^{**} = \phi^* - \Delta t A_y \phi^n$$

$$\left(I - \frac{\alpha \Delta t}{2} A_z\right)\phi^{n+1} = \phi^{**} - \frac{\Delta t}{2} A_z \phi^n.$$

What is the order of accuracy of this scheme?

7. Consider the convection–diffusion equation

$$\frac{\partial T}{\partial t} + u\frac{\partial T}{\partial x} = \alpha\frac{\partial^2 T}{\partial x^2} \qquad 0 \leq x \leq 1,$$

with the boundary conditions

$$T(0, t) = 0 \quad T(1, t) = 0.$$

This equation describes propagation and diffusion of a scalar such as temperature or a contaminant in, say, a pipe. Assume that the fluid is moving with a constant velocity u in the x direction. For the diffusion coefficient $\alpha = 0$, the solution consists of pure convection and the initial disturbance simply propagates downstream. With non-zero α, propagation is accompanied by broadening and damping.

Part 1. Pure convection ($\alpha = 0$)

Consider the following initial profile

$$T(x, 0) = \begin{cases} 1 - (10x - 1)^2 & \text{for } 0 \le x \le 0.2, \\ 0 & \text{for } 0.2 < x \le 1. \end{cases}$$

Let $u = 0.08$. The exact solution is

$$T(x, t) = \begin{cases} 1 - [10(x - ut) - 1]^2 & \text{for } 0 \le (x - ut) \le 0.2, \\ 0 & \text{for } 0.2 < (x - ut) \le 1. \end{cases}$$

(a) Solve the problem for $0 < t \le 8$ using

(i) Explicit Euler time advancement and the second-order central difference for the spatial derivative.

(ii) Leapfrog time advancement and the second-order central difference for the spatial derivative.

Plot the numerical and exact solutions for $t = 0,\ 4,\ 8$. You probably need *at least* 51 points in the x direction to resolve the disturbance. Discuss your solutions and the computational parameters that you have chosen in terms of what you know about the stability and accuracy of these schemes. Try several appropriate values for $u\Delta t/\Delta x$.

(b) Suppose u was a function of x:

$$u(x) = 0.2 \sin \pi x.$$

In this case, how would you select your time step in (a)(ii)?

(c) With the results in part (a)(i) as the motivation, the following scheme, which is known as the Lax–Wendroff scheme, has been suggested for the solution of the pure convection problem

$$T_j^{n+1} = T_j^n - \frac{\gamma}{2}\left(T_{j+1}^n - T_{j-1}^n\right) + \frac{\gamma^2}{2}\left(T_{j+1}^n - 2T_j^n + T_{j-1}^n\right),$$

where $\gamma = u\Delta t/\Delta x$. What are the accuracy and stability characteristics of this scheme? Repeat part (a)(i) with the Lax–Wendroff scheme using $\gamma = 0.8,\ 1$, and 1.1. Discuss your results using the modified equation analysis.

Part 2. Convection–diffusion.

Let $\alpha = 0.001$.

(d) Using the same initial and boundary conditions as in *Part 1*, solve the convection–diffusion equation. Repeat part (a)(i) and (ii) with the addition of the second-order central difference for the diffusion term. Discuss your results and your choices for time steps. How has the presence of diffusion term affected the physical behavior of the solution and stability properties of the numerical solutions?

(e) Suppose in the numerical formulation using leapfrog the diffusion term is lagged in time; that is, it is evaluated at step $n-1$ rather than n. Obtain the numerical solution with this scheme. Consider different values of $\alpha\Delta t/\Delta x^2$ in the range 0 to 1, and discuss your results.

8. Consider the two-dimensional Burgers equation, which is a non-linear model of the convection–diffusion process

$$\frac{\partial u}{\partial t}+u\frac{\partial u}{\partial x}+v\frac{\partial u}{\partial y}=\nu\left(\frac{\partial^2 u}{\partial x^2}+\frac{\partial^2 u}{\partial y^2}\right)$$

$$\frac{\partial v}{\partial t}+u\frac{\partial v}{\partial x}+v\frac{\partial v}{\partial y}=\nu\left(\frac{\partial^2 v}{\partial x^2}+\frac{\partial^2 v}{\partial y^2}\right).$$

We are interested in the steady state solution in the unit square, $0\le x\le 1,\ 0\le y\le 1$ with the following boundary conditions

$$\begin{aligned} &u(0,y)=u(1,y)=v(x,1)=0, &&v(x,0)=1\\ &u(x,0)=u(x,1)=\sin 2\pi x, &&v(0,y)=v(1,y)=1-y. \end{aligned}$$

The solutions of the Burgers equation usually develop steep gradients like those encountered in shock waves. Let $\nu=0.015$.

(a) Solve this problem using an explicit method. Integrate the equations until steady state is achieved (to plotting accuracy). Plot the steady state velocities u, v. (If you have access to a surface plotter such as in MATLAB, use it. If not, plot the velocities along the two lines: $x=0.5$ and $y=0.5$.) Make sure that you can stand behind the accuracy of your solution. Note that since we seek only the steady state solution, the choice of the initial condition should be irrelevant.

(b) Formulate the problem using a second-order ADI scheme for the diffusion terms and an explicit scheme for the convection terms. Give the details including the matrices involved.

9. Consider the convection–diffusion equation

$$\begin{aligned} &u_t+cu_x=\alpha u_{xx} \quad 0\le x\le 1\\ &u(x,0)=\exp\left[-200(x-0.25)^2\right] \quad u(0,t)=0. \end{aligned}$$

Take $\alpha=0$ and $c=1$ and solve using second-order central differences in x and Euler and fourth-order Runge–Kutta time advancements. Predict and verify the maximum Δt for each of these schemes. Repeat using upwind second-order spatial differences. How would the stability constraints change for non-zero α (e.g., $\alpha=0.1$)? Plot solutions at $t=0, 0.5, 1$.

10. Seismic imaging is being used in a wide variety of applications from oil exploration to non-intrusive medical observations. We want to numerically examine a one-dimensional model of a seismic imaging problem to see the effects that variable sound speeds between different media have on the transmission and reflection of an acoustic wave. The equation we will consider is the one-dimensional homogeneous scalar wave equation:

$$\frac{\partial^2 u}{\partial t^2} - c^2(x)\frac{\partial^2 u}{\partial x^2} = 0 \quad t \geq 0, \quad -\infty < x < \infty, \tag{1}$$

with initial conditions

$$u(x,0) = u_o(x) \quad u_t(x,0) = 0$$

where $c > 0$ is the speed of sound. The x domain for this problem is infinite. To cope with this numerically we truncate the domain to $0 \leq x \leq 4$. However, to do this we need to specify some conditions at the domain edges $x = 0$ and $x = 4$ such that computed waves will travel smoothly out of the computational domain as if it extended to infinity. A "radiation condition" (the Sommerfeld radiation condition) would specify that at ∞ all waves are outgoing, which is necessary for the problem to be well posed. In one-dimensional problems, this condition may be exactly applied at a finite x: we want only outgoing waves to be supported at our domain edges. That is, at $x = 4$ we want our numerical solution to support only right-going waves and at $x = 0$ we want it to support only left-going waves. If we factor the operators in the wave equation we will see more explicitly what must be done

$$\left(\frac{\partial}{\partial t} - c\frac{\partial}{\partial x}\right)\left(\frac{\partial}{\partial t} + c\frac{\partial}{\partial x}\right) u = 0. \tag{2}$$

The right-going portion of the solution is

$$\left(\frac{\partial}{\partial t} + c\frac{\partial}{\partial x}\right) u = 0 \tag{3}$$

and the left-going portion of the solution is

$$\left(\frac{\partial}{\partial t} - c\frac{\partial}{\partial x}\right) u = 0. \tag{4}$$

So at $x = 4$ we need to solve equation (3) rather than equation (1) to ensure only an outgoing (right-going) solution. Likewise, at $x = 0$ we will solve equation (4) rather than equation (1).

For time advancement it is recommended that equation (1) be broken into two first-order equations in time:

$$\frac{\partial u_1}{\partial t} = u_2 \qquad \text{and} \qquad \frac{\partial u_2}{\partial t} = c^2(x)\frac{\partial^2 u_1}{\partial x^2}.$$

The boundary conditions become

$$\left.\frac{\partial u_1}{\partial t}\right|_{x=0} = c(0)\left.\frac{\partial u_1}{\partial x}\right|_{x=0} \qquad \left.\frac{\partial u_1}{\partial t}\right|_{x=4} = -c(4)\left.\frac{\partial u_1}{\partial x}\right|_{x=4}.$$

Second-order differencing is recommended for the spatial derivative (first order at the boundaries). This problem requires high accuracy for the solution and you will find that at least $N = 400$ points should be used. Compare a solution with fewer points to the one you consider to be accurate. Use an accurate method for time advancement; fourth-order Runge–Kutta is recommended. What value of c should be used for an estimate of the maximum allowable Δt for stable solution? *Estimate* the maximum allowable time step via a modified wavenumber analysis. Take $u(x, t = 0) = \exp[-200(x - 0.25)^2]$ and specify $c(x)$ as follows:

(a) Porous sandstone: $c(x) = 1$.
(b) Transition to impermeable sandstone: $c(x) = 1.25 - 0.25 \tanh[40(0.75 - x)]$.
(c) Impermeable sandstone: $c(x) = 1.5$.
(d) Entombed alien spacecraft: $c(x) = 1.5 - \exp[-300(x - 1.75)^2]$.

Plot $u(x)$ for several ($\sim$8) different times in the calculation as a wave is allowed to propagate through the entire domain.

11. Consider a two-dimensional convection–diffusion equation

$$\frac{\partial \Theta}{\partial t} + U(x, y)\frac{\partial \Theta}{\partial x} + V(x, y)\frac{\partial \Theta}{\partial y} = \alpha\left(\frac{\partial^2 \Theta}{\partial x^2} + \frac{\partial^2 \Theta}{\partial y^2}\right),$$

where $-1 \leq y \leq 1$ and $0 \leq x \leq 10$. This equation may be used to model thermal entry problems where a hot fluid is entering a rectangular duct with a cold wall and an insulated wall. Appropriate boundary conditions for such a problem are shown in the following figure. Set up the problem using a second-order approximate factorization technique. Discuss the advantages of this technique over explicit and unfactored implicit methods.

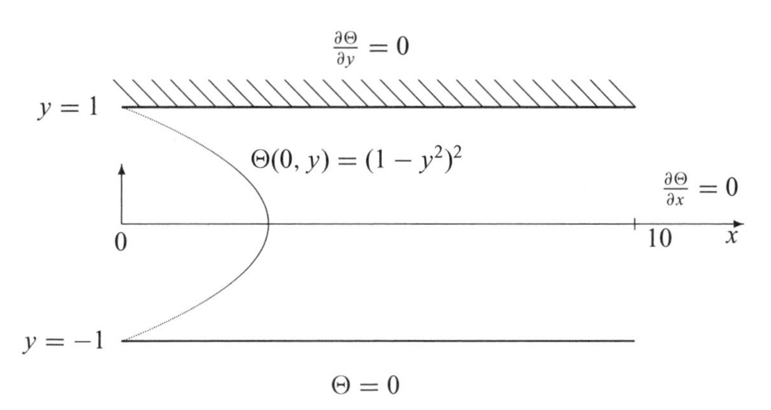

12. Consider the convection equation

$$\frac{\partial T}{\partial t} + u\frac{\partial T}{\partial x} = 0 \quad 0 \leq x \leq 10,$$

with the boundary condition

$$T(0, t) = 0.$$

This equation describes the pure convection phenomenon; i.e., an initial disturbance simply propagates downstream with the velocity u.

Consider the following initial profile

$$T(x,0)=\begin{cases}\cos^2(\pi x)-\cos(\pi x) & \text{for } 0\le x\le 2,\\ 0 & \text{for } 2<x\le 10.\end{cases}$$

The exact solution is

$$T(x,t)=\begin{cases}\cos^2[\pi(x-ut)]-\cos[\pi(x-ut)] & \text{for } 0\le (x-ut)\le 2,\\ 0 & \text{for } 2<(x-ut)\le 10.\end{cases}$$

Let $u=0.8$. Solve the problem for $0<t\le 8$ using

(a) Explicit Euler time advancement and the second-order central difference for the spatial derivative.
(b) Explicit Euler time advancement and the second-order upwind difference for the spatial derivative.
(c) Leapfrog time advancement and the second-order central difference for the spatial derivative.

Plot the numerical and the exact solutions for $t=0,4,8$. You probably need *at least* 101 points in the x direction to resolve the disturbance. Try two or three different values of $\gamma=u\Delta t/\Delta x$. Compare and discuss your solutions and the computational parameters that you have chosen in terms of what you know about the stability and accuracy of these schemes.

For method (c), perform the modified equation analysis and solve the equation with the value of $\gamma=1$ using second-order Runge–Kutta method for the start-up step. Discuss your results.

13. The heat equation with a source term is

$$\frac{\partial T}{\partial t}=\alpha\frac{\partial^2 T}{\partial x^2}+S(x)\quad 0\le x\le L_x.$$

The initial and boundary conditions are

$$T(x,0)=0\quad T(0,t)=0\quad T(L_x,t)=T_{\text{steady}}(L_x).$$

Take $\alpha=1$, $L_x=15$, and $S(x)=-(x^2-4x+2)e^{-x}$. The exact steady solution is

$$T_{\text{steady}}(x)=x^2e^{-x}.$$

(a) Verify that $T_{\text{steady}}(x)$ is indeed the exact steady solution. Plot $T_{\text{steady}}(x)$.
(b) Using explicit Euler for time advancement and the second-order central difference scheme for the spatial derivative, solve the equation to steady state on a uniform grid. Plot the exact and numerical steady solutions for $N_x=10,20$.
(c) Repeat your calculations using the non-uniform grid $x_j=L_x[1-\cos(\frac{\pi j}{2N_x})]$, $j=0,\ldots,N_x$ and an appropriate finite difference scheme for non-uniform grid.
(d) Transform the differential equation to a new coordinate system using the transformation

$$\zeta=\cos^{-1}\left(1-\frac{x}{L_x}\right).$$

Solve the resulting equation to the steady state and plot the exact and numerical steady solutions for $N_x = 10, 20$.

(e) Repeat (c) using the Crank–Nicolson method for time advancement. Show that you can take fewer time steps to reach steady state.

For each method, find the maximum time step required for stable solution. Also, for each method with $N_x = 20$, plot the transient solutions at two intermediate times, e.g., at $t = 2$ and $t = 10$. Compare and discuss all results obtained in terms of accuracy and stability. Compare the number of time steps required for each method to reach steady state.

14. The forced convection–diffusion equation

$$\frac{\partial \phi}{\partial t} - u\frac{\partial \phi}{\partial x} = \alpha \frac{\partial^2 \phi}{\partial x^2} + S(x) \qquad 0 \le x \le 1$$

has the following boundary conditions:

$$\phi(0,t) = 0 \qquad \frac{\partial \phi}{\partial x}(1,t) = 1.$$

(a) We would like to use the explicit Euler in time and the second-order central difference in space to solve this equation numerically. Using matrix stability analysis, find the stability condition of this method for arbitrary combinations of u, α, and Δx. Note that u and α are positive constants. What is the stability condition for $\Delta x << 1$ (i.e., Δx is much less than 1)?

(b) Let $\alpha = 0$, $u = 1$, and $S(x) = 0$. Suppose we use fourth-order Padé scheme for the spatial derivative and one of the following schemes for the time advancement:

(i) Explicit Euler
(ii) Leapfrog
(iii) Fourth-order Runge–Kutta

Based on what you know about these schemes obtain the maximum time step for stability. *Hint:* Although the matrix stability analysis is probably the easiest method to use in (a), it may not be the easiest for (b).

(c) How would you find the maximum time step in (b) if instead of $u = 1$ you had $u = \sin \pi x$?

15. The well-known non-linear Burgers equation is

$$\frac{\partial u}{\partial t} + u\frac{\partial u}{\partial x} = \alpha \frac{\partial^2 u}{\partial x^2} \qquad 0 \le x \le 1.$$

The boundary conditions are

$$u(0,t) = 0 \quad u(1,t) = 0.$$

We would like to solve this problem using an implicit second-order method in time and a second-order method in space. Write down the discrete form of the equation. Develop an algorithm for the solution of this equation. Show how you can avoid iterations in your algorithm. Give all the details including matrices involved.

16. The following iterative scheme is used to solve $A\boldsymbol{x} = \boldsymbol{b}$:

$$\boldsymbol{x}^{(k+1)} = (I + \alpha A)\boldsymbol{x}^{(k)} - \alpha \boldsymbol{b},$$

where α is a real constant, and A is the following tridiagonal matrix that has resulted from a finite difference approximation:

$$A = \begin{bmatrix} -2 & 1 & & & \\ 1 & -2 & 1 & & \\ & & \ddots & \ddots & \ddots \\ & & & 1 & -2 \end{bmatrix}.$$

Under what conditions on α does this algorithm converge?

17. The following is a 1D boundary value problem:

$$\frac{d^2u}{dx^2} + \alpha \frac{du}{dx} + \beta u = f(x)$$
$$u(0) = u_o \quad u(L) = u_L.$$

(a) Set up the system of equations required to solve this boundary value problem directly using second-order central differences.
(b) Suppose we wish to use the Point–Jacobi method to solve this system. With $\beta(\Delta x)^2 = 3$, state the conditions on $\alpha \Delta x$ necessary for convergence.
(c) Approximately how many iterations are necessary to reduce the error to 0.1% of its original value for $\beta(\Delta x)^2 = 3$ and $\alpha \Delta x = 1.75$?
(d) If a shooting method were to be used, how many shots would be necessary to solve this problem?

18. The equation $A\boldsymbol{x} = \boldsymbol{f}$ is solved using two iterative schemes of the form

$$A_1 \boldsymbol{x}^{(k+1)} = A_2 \boldsymbol{x}^{(k)} + \boldsymbol{f},$$

where

$$A = \begin{bmatrix} a & b \\ c & d \end{bmatrix} \quad \text{and} \quad A_1 - A_2 = A.$$

The two schemes are given by

$$\text{(i) } A_1 = \begin{bmatrix} a & 0 \\ 0 & d \end{bmatrix} \qquad \text{(ii) } A_1 = \begin{bmatrix} a & b \\ 0 & d \end{bmatrix}.$$

What is the condition among the elements of A so that both schemes would converge? Compare the convergence rates of the two schemes.

19. The steady state temperature distribution $u(x, y)$ in the rectangular copper plate

below satisfies Laplace's equation:

$$\frac{\partial^2 u}{\partial x^2} + \frac{\partial^2 u}{\partial y^2} = 0.$$

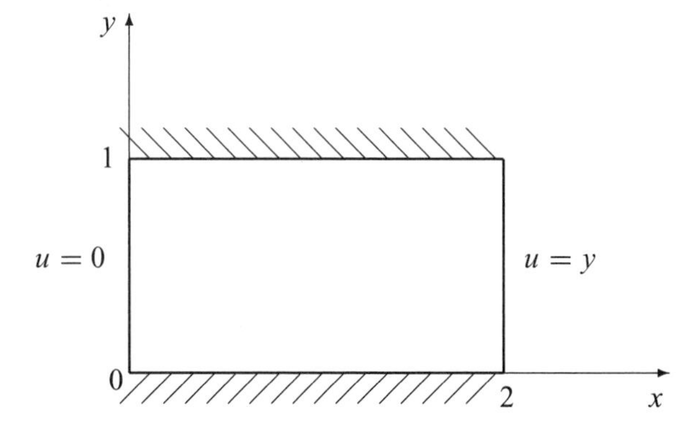

The upper and lower boundaries are perfectly insulated; the left side is kept at $0°C$, and the right side at $f(y) = y°C$. The exact solution can be obtained analytically using the method of separation of variables and is given by

$$u(x, y) = \frac{x}{4} - 4 \sum_{\substack{n=1 \\ n \text{ odd}}}^{\infty} \frac{1}{(n\pi)^2 \sinh 2n\pi} \sinh n\pi x \cos n\pi y.$$

In this exercise we will find numerical approximations to the steady state solution.

(a) First write a program to compute the steady state solution to the second-order finite difference approximation of the heat equation using the Jacobi iteration method. You should use N_x and N_y uniformly spaced points in the horizontal and vertical directions, respectively (this includes the points on the boundaries).

(b) Now with $N_x = 11$ and $N_y = 11$ apply the Jacobi iteration to the discrete equations until the solution reaches steady state. To start the iterations, initialize the array with zeroes except for the boundary elements corresponding to $u = y$.

You can monitor the progress of the solution by watching the value of the solution at the center of the plate: $(x, y) = (1, 0.5)$. How many iterations are required until the solution at (1, 0.5) *steadily* varies by no more than 0.00005 between iterations? At this point, how does the numerical approximation compare to the analytical solution? What is the absolute error? What is the error in the numerical approximation relative to the analytical solution (percentage error)?

Plot isotherms of the numerical and exact temperature distributions (say, 16 isotherms). Use different line styles for the numerical and analytical isotherms and put them on the same axes, but be sure to use the same temperature values for each set of isotherms (that is, the same contour levels).

Repeat the same steps above with $N_x = 21$ and $N_y = 21$.

(c) Repeat (b) using the Gauss–Seidel iteration and SOR. Compare the performance of the methods.

FURTHER READING

Ames, W. F. *Numerical Methods for Partial Differential Equations*, Third Edition. Academic Press, 1992.

Briggs, W. L. *A Multigrid Tutorial*. Society for Industrial and Applied Mathematics (SIAM), 1987.

Ferziger, J. H. *Numerical Methods for Engineering Application*, Second Edition. Wiley, 1998, Chapters 6, 7, and 8.

Greenbaum, A. *Iterative Methods for Solving Linear Systems*. Society for Industrial and Applied Mathematics (SIAM), 1997.

Lapidus, L. and Pinder, G. F. *Numerical Solution of Partial Differential Equations in Science and Engineering*. Wiley, 1982, Chapters 4, 5, and 6.

Morton, K. W. and Mayers, D. F. *Numerical Solution of Partial Differential Equations*. Cambridge University Press, 1994.

Press, W. H., Teukolsky, S. A., Vetterling, W. T., and Flannery, B. P. *Numerical Recipes: The Art of Scientific Computing*, Second Edition. Cambridge University Press, 1992, Chapter 19.

Varga, R. *Matrix Iterative Analysis*. Prentice-Hall, 1962.

Young, D. *Iterative Solution of Large Linear Systems*. Academic Press, 1971.

6

Discrete Transform Methods

Transform methods can be viewed as semi-analytical alternatives to finite differences for spatial differentiation in applications where *high degree of accuracy* is required. This chapter is an introduction to transform methods, also referred to as spectral methods, for solution of partial differential equations. We shall begin with the discrete Fourier transform, which is applied to numerical differentiation of periodic data and for solving elliptic PDEs in rectangular geometries. Discrete Fourier transform is also used extensively in signal processing, but this important application of transform methods will not be discussed here. For non-periodic data we will use transform methods based on Chebyshev polynomial expansions. Once the basic machinery for numerical differentiation with transform methods is developed, we shall see that their use for solving partial differential equations is straightforward.

6.1 Fourier Series

Consider the representation of a continuous *periodic* function f as a combination of pure harmonics

$$f(x) = \sum_{k=-\infty}^{\infty} \hat{f}_k e^{ikx}, \tag{6.1}$$

where $\hat{f}_k$ is the Fourier coefficient corresponding to the wavenumber k. Here the k values are integers because the period is taken to be 2π. In Fourier analysis one is interested in knowing what harmonics contribute to f and by how much. This information is provided by $\hat{f}_k$. The Fourier series for the derivative of $f(x)$ is obtained by simply differentiating (6.1)

$$f'(x) = \sum_{k=-\infty}^{\infty} ik\hat{f}_k e^{ikx}. \tag{6.2}$$

By analogy with the Fourier transform of f in (6.1), the Fourier coefficients of f' are $ik\hat{f}_k$. In this section the machinery for calculating $\hat{f}_k$ will be developed for *discrete data*. Once $\hat{f}_k$ is obtained, it is simply multiplied by ik to obtain the Fourier coefficients of f'. The result is then substituted in the discrete version of (6.2) to compute f'.

6.1.1 Discrete Fourier Series

If the periodic function f is defined only on a discrete set of N grid points, $x_0, x_1, x_2, \ldots, x_{N-1}$, then f can be represented by a discrete Fourier transform. Discrete Fourier transform of a sequence of N numbers, $f_0, f_1, f_2, \ldots, f_{N-1}$ is defined by

$$f_j = \sum_{k=-\frac{N}{2}}^{\frac{N}{2}-1} \hat{f}_k e^{ikx_j} \quad j = 0, 1, 2, \ldots, N-1 \qquad (6.3)$$

where

$$\hat{f}_{-\frac{N}{2}}, \hat{f}_{-\frac{N}{2}+1}, \ldots, 0, \ldots, \hat{f}_{\frac{N}{2}-1}$$

are the discrete Fourier coefficients of f. Here, we take N to be even and the period of f to be 2π. A consequence of 2π periodicity is having integer wavenumbers. The sequence f_j consists of the values of f evaluated at equidistant points along the axis $x_j = jh$ with the grid spacing $h = 2\pi/N$. Note that f is assumed to be a periodic function with $f_0 = f_N$, and thus, the sequence $f_0, f_1, \ldots, f_{N-1}$ does not involve any redundancy. In the more general case of period of length L the wavenumbers appearing in the argument of the exponential would be $(2\pi/L)k$ instead of k, and the grid spacing becomes $h = L/N$, which results in an identical expression for the arguments of the exponentials as in the 2π periodic case. Thus, the actual period does not appear in the expression for the discrete Fourier transform of f, but it does appear in the expression for its derivative (see (6.2)).

Equation (6.3) constitutes N algebraic equations for the unknown (complex) Fourier coefficients $\hat{f}_k$. However, instead of using Gauss elimination, or some other solution technique from linear algebra to solve this system, it is much easier and more efficient, to use the *discrete* orthogonality property of the Fourier series to get the Fourier coefficients. Therefore, we will first establish the discrete orthogonality of Fourier series. Consider the summation

$$I = \sum_{j=0}^{N-1} e^{ikx_j} e^{-ik'x_j} = \sum_{j=0}^{N-1} e^{ih(k-k')j}.$$

If $h(k - k')$ is not a multiple of 2π, then I is a geometric series with the multiplier

$e^{ih(k-k')}$. Thus, for $k - k' \neq mN$ (m is an integer),

$$I = \frac{1 - e^{ih(k-k')N}}{1 - e^{ih(k-k')}}.$$

Since $h = 2\pi/N$, the numerator is zero and we have the following statement of discrete orthogonality:

$$\sum_{j=0}^{N-1} e^{ikx_j} e^{-ik'x_j} = \begin{cases} N, & \text{if } k = k' + mN,\ m = 0,\ \pm 1,\ \pm 2, \dots \\ 0, & \text{otherwise.} \end{cases} \tag{6.4}$$

Now, we will use this important result to obtain the Fourier coefficients $\hat{f}_k$. Multiplying both sides of (6.3) by $e^{-ik'x_j}$ and summing from $j = 0$ to $N - 1$ results in

$$\sum_{j=0}^{N-1} f_j e^{-ik'x_j} = \sum_{k=-\frac{N}{2}}^{\frac{N}{2}-1} \sum_{j=0}^{N-1} \hat{f}_k e^{ix_j(k-k')}.$$

Using the orthogonality property (6.4), we have

$$\hat{f}_k = \frac{1}{N} \sum_{j=0}^{N-1} f_j e^{-ikx_j} \quad k = -\frac{N}{2}, -\frac{N}{2} + 1, \dots, \frac{N}{2} - 1. \tag{6.5}$$

Equations (6.3) and (6.5) constitute the discrete Fourier transform pair for the discrete data, f_j. Equation (6.5) is sometimes referred to as the forward transform (from the physical space x to the Fourier space k) and (6.3) is referred to as the inverse transform (for recovering the function from its Fourier coefficients).

6.1.2 Fast Fourier Transform

For complex data, straightforward summations for each transform ((6.3) or (6.5)) requires about $4N^2$ arithmetic operations (multiplications and additions), assuming that the values of the trigonometric functions are tabulated. An ingenious algorithm, developed in the 1960s and called the fast Fourier transform (FFT), reduces this operations count to $O(N \log_2 N)$. This is a dramatic reduction for large values of N. The original algorithm was developed for $N = 2^m$, but algorithms that allow more general values of N have since been developed. The fast Fourier transform algorithm has been the subject of many articles and books and therefore will not be presented here. Very efficient FFT computer programs are also available for virtually all computer platforms used for scientific computing. For example, *Numerical Recipes* has a set of programs for the general FFT algorithm and several of its useful variants for real functions and for sine and cosine transforms, which are mentioned later in this chapter.

6.1.3 Fourier Transform of a Real Function

Whether f is real or complex the Fourier coefficients of f are, generally complex. However, when f is real, there is a useful relationship relating its Fourier coefficients corresponding to negative and positive wavenumbers. This property reduces the storage requirements; the original N real data points f_j are equivalently represented by $N/2$ complex Fourier coefficients. We can easily derive this relationship by revisiting (6.5). Changing k to $-k$ in (6.5) produces

$$\hat{f}_{-k} = \frac{1}{N}\sum_{j=0}^{N-1} f_j e^{ikx_j}. \tag{6.6}$$

Taking the complex conjugate of this expression and noting that since f is real it is equal to its own complex conjugate, we obtain

$$\hat{f}^*_{-k} = \frac{1}{N}\sum_{j=0}^{N-1} f_j e^{-ikx_j}. \tag{6.7}$$

Comparison with (6.5) leads to this important result for real functions

$$\hat{f}_{-k} = \hat{f}^*_k. \tag{6.8}$$

As mentioned in the previous section, there are fast transform programs for real functions that take advantage of this property to reduce the required memory and execution time.

EXAMPLE 6.1 Calculation of Discrete Fourier Transform

(a) Consider the periodic function $f(x) = \cos 3x$ with period 2π, defined on the discrete set of points $x_j = (2\pi/N)j$, where $j = 0, \ldots, N-1$. Since

$$f_j = \cos 3x_j = \sum_{k=-\frac{N}{2}}^{\frac{N}{2}-1} \hat{f}_k e^{ikx_j} = \sum_{k=-\frac{N}{2}}^{\frac{N}{2}-1} \hat{f}_k(\cos kx_j + i\sin kx_j),$$

calculation of the Fourier coefficients is straightforward and obtained by inspection. They are given by

$$\hat{f}_k = \begin{cases} 1/2 & \text{if } k = \pm 3, \\ 0 & \text{otherwise.} \end{cases}$$

The result is independent of the number of discrete points N as long as $N \geq 8$.

(b) Consider now the periodic square function (Figure 6.1), which is given by

$$f(x) = \begin{cases} 1 & \text{if } 0 \leq x < \pi \\ -1 & \text{if } \pi \leq x < 2\pi, \end{cases}$$

and defined on the same discrete set of points. Let $N = 16$. Instead of directly using (6.5) to calculate the Fourier coefficients, we use *Numerical*

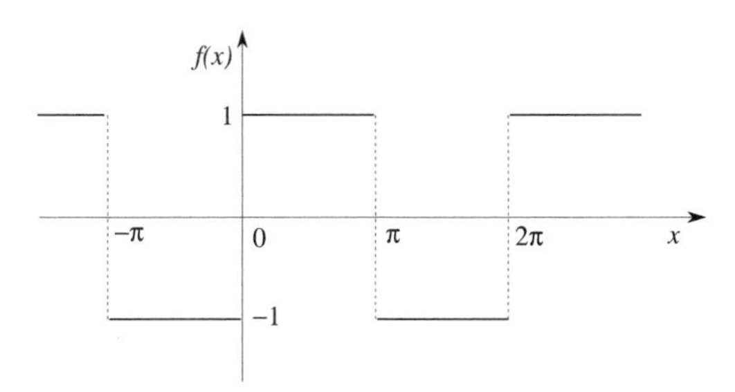

Figure 6.1 Periodic square function in Example 6.1(b).

Recipes' `realft` fast Fourier transform subroutine for real functions. The magnitudes of the Fourier coefficients are shown in Figure 6.2 and the coefficients corresponding to the positive wavenumbers are tabulated below. Fourier coefficients for negative wavenumbers are given by $\hat{f}_{-|k|} = \hat{f}^*_{|k|}$ because $f(x)$ is real.

k	Re($\hat{f}_k$)	Im($\hat{f}_k$)	$\lvert\hat{f}_k\rvert$
0	0	0	0
1	0.125	−0.628	0.641
2	0	0	0
3	0.125	−0.187	0.225
4	0	0	0
5	0.125	−0.084	0.150
6	0	0	0
7	0.125	−0.025	0.127
8	0	0	0

Using (6.5), it can be shown that if f_j is an odd function then its discrete

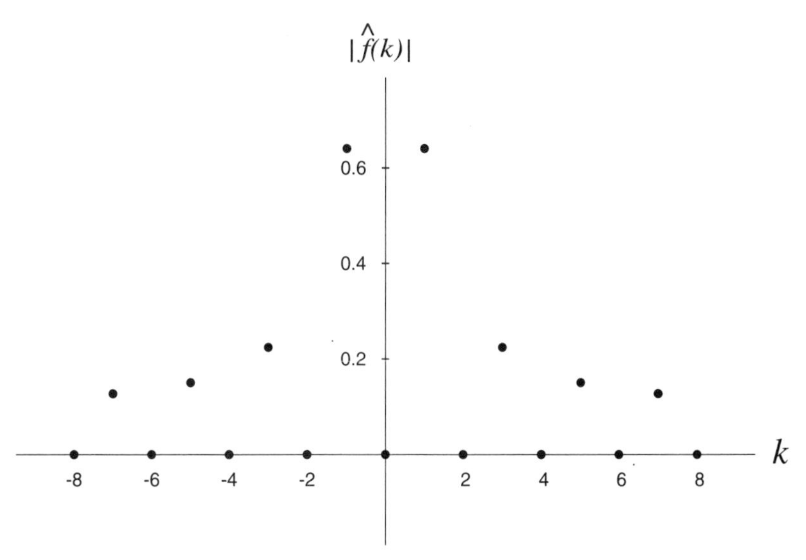

Figure 6.2 The magnitudes of the Fourier coefficients of the square function in Example 6.1(b).

Fourier transform $\hat{f}_k$ is imaginary and odd. The square function in this example can be made odd by redefining its values at 0 and π to be zeros instead of 1 and -1. In this case, the real part of the Fourier coefficients would be zero and the imaginary part would be unaltered compared to the original case.

6.1.4 Discrete Fourier Series in Higher Dimensions

The results and methodology of discrete Fourier transform can be extended to multiple dimensions in a straightforward manner. Consider the function $f(x, y)$ which is doubly periodic in the x and y directions and discretized using N_1 grid points in x and N_2 grid points in y. The two-dimensional Fourier series representation of f is given by

$$f(x_m, y_l) = \sum_{k_1=-\frac{N_1}{2}}^{\frac{N_1}{2}-1} \sum_{k_2=-\frac{N_2}{2}}^{\frac{N_2}{2}-1} \hat{f}_{k_1,k_2} e^{ik_1x_m} e^{ik_2y_l}$$

$$m = 0, 1, 2, \ldots, N_1 - 1 \quad l = 0, 1, 2, \ldots, N_2 - 1, \tag{6.9}$$

where $\hat{f}$ is the (complex) Fourier coefficient of f corresponding to wavenumbers k_1 and k_2 in the x and y directions respectively. Using the orthogonality result (6.4) for each direction, we obtain

$$\hat{f}_{k_1,k_2} = \frac{1}{N_1}\frac{1}{N_2}\sum_{m=0}^{N_1}\sum_{l=0}^{N_2} f_{m,l} e^{-ik_1x_m} e^{-ik_2y_l} \tag{6.10}$$

$$k_1 = -\frac{N_1}{2}, \; -\frac{N_1}{2}+1, \ldots, \frac{N_1}{2}-1 \quad \text{and} \quad k_2 = -\frac{N_2}{2}, \; -\frac{N_2}{2}+1, \ldots, \frac{N_2}{2}-1.$$

If f is real, it can be easily shown as in the previous section that

$$\hat{f}^*_{-k_1,-k_2} = \hat{f}_{k_1,k_2}.$$

Thus, Fourier coefficients in one half (*not one quarter*) of the (k_1, k_2) space are sufficient to determine all the Fourier coefficients in the entire (k_1, k_2) plane. All these results can be generalized to higher dimensions. For example, in three dimensions

$$\hat{f}^*_{-\boldsymbol{k}} = \hat{f}_{\boldsymbol{k}}$$

where $\boldsymbol{k} = (k_1, \; k_2, \; k_3)$ is the wavenumber vector.

6.1.5 Discrete Fourier Transform of a Product of Two Functions

The following is an important result that will be used later for the solution of non-linear equations by transform methods. Let

$$H(x) = f(x)g(x).$$

Our objective is to express the Fourier transform of H in terms of the Fourier transforms of f and g. The discrete Fourier transform of H is

$$\hat{H}_m = (\widehat{fg})_m = \frac{1}{N}\sum_{j=0}^{N-1} f_j g_j e^{-imx_j}.$$

Substituting for f_j and g_j their respective Fourier representations, we obtain

$$\hat{H}_m = \frac{1}{N}\sum_{j=0}^{N-1}\sum_{k}\sum_{k'} \hat{f}_k \hat{g}_{k'} e^{ikx_j} e^{ik'x_j} e^{-imx_j}. \tag{6.11}$$

The sum over j is non-zero only if $k + k' = m$ or $m \pm N$ (recall that, $x_j = (2\pi N)j$). The part of the summation corresponding to $k + k' = m \pm N$ is known as the *aliasing error* and should be discarded because the Fourier exponentials corresponding to these wavenumbers cannot be resolved on the grid of size N. Thus, using the definition (6.5) the Fourier transform of the product is

$$\hat{H}_m = \sum_{k=-N/2}^{N/2-1} \hat{f}_k \hat{g}_{m-k}. \tag{6.12}$$

This is the convolution sum of the Fourier coefficients of f and g. The inverse transform of $\hat{H}_m$ is sometimes used as the means to calculate the product of f and g. If we simply multiplied f and g at each grid point, the resulting discrete function would be "contaminated" by the aliasing errors and would not be equal to the inverse transform of $\hat{H}_m$ in (6.12). Aliasing errors are simply ignored in many calculations, in part because the alternative, alias-free method of calculation of the product via (6.12) is expensive, requiring $O(N^2)$ operations, and aliasing errors are usually small if sufficient number of grid points are used. However, in some large-scale computations aliasing errors have led to very inaccurate solutions. We will illustrate the effect of aliasing error in the following example.

EXAMPLE 6.2 Discrete Fourier Transform of a Product–Aliasing

Consider the functions $f(x) = \sin 2x$ and $g(x) = \sin 3x$ defined on the grid points $x_j = (2\pi/N)j$, where $j = 0, \ldots, N-1$. For $N \geq 8$, their discrete Fourier transforms are

$$\hat{f}_k = \begin{cases} \mp i/2 & \text{if } k = \pm 2 \\ 0 & \text{otherwise,} \end{cases} \quad \text{and} \quad \hat{g}_k = \begin{cases} \mp i/2 & \text{if } k = \pm 3 \\ 0 & \text{otherwise.} \end{cases}$$

Using trigonometric identities, their product $H(x) = f(x)g(x)$ is equal to $0.5(\cos x - \cos 5x)$.

We want to calculate the discrete Fourier transform of $H(x)$ using discrete values of f and g. For $N = 16$, using (6.12) or simply multiplying f and g at each grid point and inverse transforming, we obtain

$$\hat{H}_k = \begin{cases} 1/4 & \text{if } k = \pm 1 \\ -1/4 & \text{if } k = \pm 5 \\ 0 & \text{otherwise,} \end{cases}$$

which is the Fourier transform of the discrete function $0.5(\cos x_j - \cos 5x_j)$. Thus the exact Fourier coefficients of $H(x)$ are recovered.

We use now a smaller number of points ($N = 8$) to calculate the discrete Fourier coefficients of $H(x)$. Equation (6.12) gives

$$\hat{H}_k = \begin{cases} 1/4 & \text{if } k = \pm 1 \\ 0 & \text{otherwise,} \end{cases}$$

which corresponds to the discrete function $0.5 \cos x_j$. The 8-point grid is able to resolve Fourier modes up to the wavenumber $k = N/2 = 4$. Therefore, the part of $H(x)$ corresponding to $k = 5$ is lost when representing $H(x)$ discretely. The error involved is the truncation error since it results from truncating the Fourier series.

If we multiply f and g at each grid point and Fourier transform the result, we obtain

$$\hat{H}_k = \begin{cases} 1/4 & \text{if } k = \pm 1 \\ -1/4 & \text{if } k = \pm 3 \\ 0 & \text{otherwise,} \end{cases}$$

which is the Fourier transform of the discrete function $0.5(\cos x_j - \cos 3x_j)$! We notice the appearance of a new mode: $\cos 3x_j$. This is the aliasing error that has contaminated the results. It is the alias or misrepresentation of the $\cos 5x$ mode that appears (uses the alias) as $\cos 3x$. This is illustrated in Figure 6.3.

6.1.6 Discrete Sine and Cosine Transforms

If the function f is not periodic, transforms based on other than harmonic functions are usually more suitable representations of f. For example, if f is an even function (i.e., $f(x) = f(-x)$), expansion based on cosines would be a more suitable representation for f.

Consider the function f defined on an equidistant set of $N + 1$ points on the interval $0 \leq x \leq \pi$ on the real axis. Discrete cosine transform of f is defined

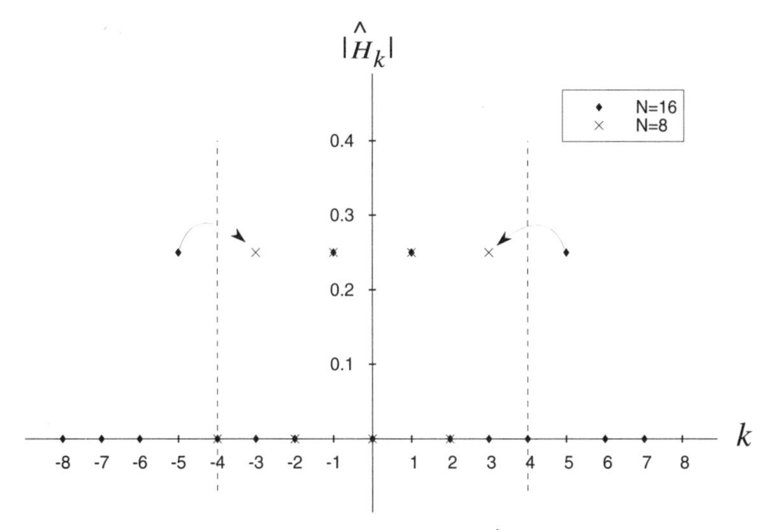

Figure 6.3 The magnitude of the Fourier coefficient $\hat{H}_k$ for $N = 8$ and $N = 6$ in Example 6.2.

by the following pair of relations

$$f_j = \sum_{k=0}^{N} a_k \cos kx_j \quad j = 0, 1, 2, \ldots, N \tag{6.13}$$

$$a_k = \frac{2}{c_k N} \sum_{j=0}^{N} \frac{1}{c_j} f_j \cos kx_j \quad k = 0, 1, 2, \ldots, N, \tag{6.14}$$

where

$$c_l = \begin{cases} 2 & \text{if } l = 0, N \\ 1 & \text{otherwise,} \end{cases}$$

and $x_j = jh$ with $h = \pi/N$. Note that in contrast to the periodic Fourier transform, the values of f at both ends of the interval, f_0 and f_N, are included. Relation (6.13) is the definition of cosine transform for f. As in Fourier transforms, (6.14) is derived using the discrete orthogonality property of the cosines:

$$\sum_{j=0}^{N} \frac{1}{c_j} \cos kx_j \cos k'x_j = \begin{cases} 0 & \text{if } k \neq k' \\ \frac{1}{2} c_k N & \text{if } k = k'. \end{cases} \tag{6.15}$$

Discrete orthogonality of cosines given in (6.15) can be easily derived by substituting complex exponential representations for cosines in (6.15) and using geometric series, as was done in the Fourier case. Derivation of both equations (6.14) and (6.15) are left as exercises at the end of this chapter. Similarly, if f is an odd function (i.e., $f(x) = -f(-x)$), then it is best represented based on sine series. The sine transform pair is given by

$$f_j = \sum_{k=0}^{N} b_k \sin kx_j \quad j = 0, 1, 2, \ldots, N \tag{6.16}$$

$$b_k = \frac{2}{N} \sum_{j=0}^{N} f_j \sin kx_j \quad k = 0, 1, 2, \ldots, N. \tag{6.17}$$

Note that the $\sin kx_j$ term is zero at both ends of the summation index; they are included here to maintain similarity with the cosine transform relations.

EXAMPLE 6.3 Calculation of the Discrete Sine and Cosine Transforms

Consider the function $f(x) = x^2/\pi^2$, defined on the discrete points $x_j = (\pi/N)j$, where $j = 0, \ldots, N$. Let $N = 16$. We use *Numerical Recipes'* `cosft1` and `sinft` which are fast cosine and sine transform routines. The magnitudes of the coefficients are plotted in Figure 6.4. It is clear that the coefficients of the cosine expansion decay faster than those of the sine expansion. The sine expansion needs more terms to approximate the function on the whole interval as accurately as the cosine approximation because $f(\pi) \neq 0$. The odd periodic continuation of $f(x)$ is discontinuous at $x = \pi \pm 2n\pi$, n integer; the even continuation is not discontinuous (its slope is). The discontinuity slows the convergence of the expansion.

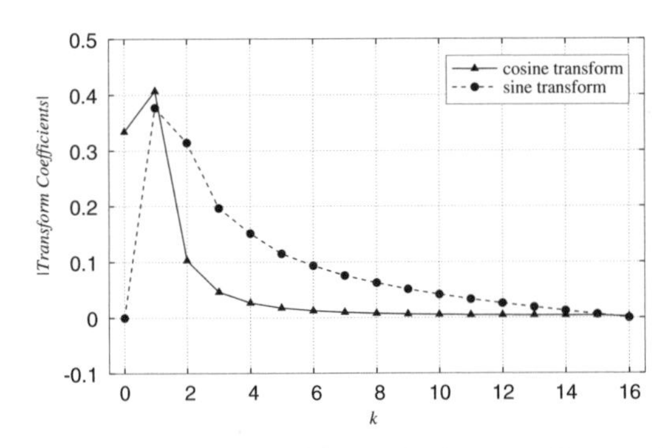

Figure 6.4 Magnitude of the cosine and sine transform coefficients for $f(x) = x^2/\pi^2$ in Example 6.3.

6.2 Applications of Discrete Fourier Series

6.2.1 Direct Solution of Finite Differenced Elliptic Equations

In this section we will give an example of a novel application of transform methods for solving elliptic partial differential equations. Consider the Poisson equation

$$\frac{\partial^2 \phi}{\partial x^2} + \frac{\partial^2 \phi}{\partial y^2} = Q(x, y)$$

with $\phi = 0$ on the boundaries of a rectangular domain. Suppose we seek a finite difference solution of this equation using a second-order finite difference scheme with $M + 1$ points in the x direction (including the boundaries) and $N + 1$ points in the y direction. Let the uniform mesh spacing in the x direction be denoted by Δ_1 and the mesh spacing in the y direction by Δ_2. The finite difference equations are

$$\phi_{i+1,j} - 2\phi_{i,j} + \phi_{i-1,j} + \frac{\Delta_1^2}{\Delta_2^2}(\phi_{i,j+1} - 2\phi_{i,j} + \phi_{i,j-1}) = \Delta_1^2 Q_{i,j}, \qquad (6.18)$$

where

$$i = 1, 2, \ldots, M - 1 \quad \text{and} \quad j = 1, 2, \ldots, N - 1$$

are the mesh points inside the domain. This is a system of linear algebraic equations for the $(N - 1) \times (M - 1)$ unknowns. As pointed out in Section 5.10, for typical values of M and N, this system of equations is usually too large for a straightforward application of Gauss elimination. Here, we shall use Fourier sine series and the fast Fourier transform algorithm to obtain the solution of this *system of algebraic equations*.

Assume a solution of the form

$$\phi_{i,j} = \sum_{k=1}^{M-1} \hat{\phi}_{k,j} \sin\left[\frac{\pi k i}{M}\right] \quad i = 1, 2, \ldots, M - 1, \ j = 1, 2, \ldots, N - 1. \qquad (6.19)$$

Whether this assumed solution would work will be determined after substitution into (6.18). Note that the assumed solution does not include the boundaries, but it is consistent with the homogeneous boundary conditions. The Fourier transform of the right-hand side is similarly expressed as

$$Q_{i,j} = \sum_{k=1}^{M-1} \hat{Q}_{k,j} \sin\left[\frac{\pi k i}{M}\right] \quad i = 1, 2, \ldots, M - 1, \quad j = 1, 2, \ldots, N - 1.$$

Substituting these representations in the finite differenced equation (6.18), we obtain

$$\sum_{k=1}^{M-1} \hat{\phi}_{k,j} \left\{ \sin\left[\frac{\pi k(i+1)}{M}\right] - 2\sin\left[\frac{\pi k i}{M}\right] + \sin\left[\frac{\pi k(i-1)}{M}\right] \right\}$$
$$+ \sum_{k=1}^{M-1} \left(\frac{\Delta_1^2}{\Delta_2^2}\right) \left\{ \hat{\phi}_{k,j+1} - 2\hat{\phi}_{k,j} + \hat{\phi}_{k,j-1} \right\} \sin\left[\frac{\pi k i}{M}\right]$$
$$= \Delta_1^2 \sum_{k=1}^{M-1} \hat{Q}_{k,j} \sin\left[\frac{\pi k i}{M}\right]. \qquad (6.20)$$

Using trigonometric identities, we have

$$\sin\left[\frac{\pi k(i+1)}{M}\right] - 2\sin\left[\frac{\pi k i}{M}\right] + \sin\left[\frac{\pi k(i-1)}{M}\right]$$
$$= \sin\left[\frac{\pi k i}{M}\right]\left[2\cos\frac{\pi k}{M} - 2\right].$$

By equating the coefficients of $\sin \pi k i/M$ in (6.20) (which amounts to using the discrete orthogonality property of the sines), we will obtain the following equation for the coefficients of the sine series:

$$\hat{\phi}_{k,j+1} + \left[\frac{\Delta_2^2}{\Delta_1^2}\left(2\cos\frac{\pi k}{M} - 2\right) - 2\right]\hat{\phi}_{k,j} + \hat{\phi}_{k,j-1} = \Delta_2^2 \hat{Q}_{k,j}. \quad (6.21)$$

For each k, this is a tridiagonal system of equations that can be easily solved.

Thus, the procedure for solving the Poisson equation can be summarized as follows. First, for each $j = 1, 2, \ldots, (N-1)$ the right-hand side function, $Q_{i,j}$ in (6.18), is Fourier sine transformed to obtain $\hat{Q}_{k,j}$:

$$\hat{Q}_{k,j} = \frac{2}{M}\sum_{i=1}^{M-1} Q_{i,j}\sin\left[\frac{\pi k i}{M}\right] \quad k = 1, 2, \ldots, M-1, \quad j = 1, 2, \ldots, N-1.$$

Then, the tridiagonal system of equations (6.21) is solved for each $k = 1, 2, \ldots, (M-1)$. Finally, ϕ_{ij} is obtained from (6.19) using discrete fast sine transform.

Thus, the two-dimensional problem has been separated into $(M-1)$ one-dimensional problems. Since each sine transform requires $O(M \log_2 M)$ operations and each tridiagonal system $O(N)$ operations, overall, the method requires $O(NM \log_2 M)$ operations. It is a direct and a low-cost method for elliptic equations. However, the class of problems for which it works is limited. One must have a uniform mesh in the direction of transform (in this case, the x direction) and the coefficients in the PDE may not be a function of the transform direction. Non-uniform meshes and non-constant coefficients may be used in the other direction(s).

It should be emphasized that this solution procedure is simply a method for solving the system of linear equations (6.18). It is not a spectral numerical solution of the Poisson equation. Spectral methods is the subject of the remaining sections of this chapter. Furthermore, the sine series only involves the interior points. However, the fact that the representation for ϕ is also consistent with the boundary conditions is a key to the success of the method. For non-homogeneous boundary conditions, a change of variables must be introduced which would transform the inhomogeneity to the right-hand side term. For problems with Neumann boundary conditions, cosine series can be used instead of sine series.

EXAMPLE 6.4 Poisson Equation With Non-homogeneous Boundary Conditions

Consider the Poisson equation

$$\frac{\partial^2 \psi}{\partial x^2} + \frac{\partial^2 \psi}{\partial y^2} = 30(x^2 - x) + 30(y^2 - y) \quad 0 \le x \le 1, \quad 0 \le y \le 1,$$

with $\psi(0, y) = \sin 2\pi y$ and $\psi = 0$ on the other boundaries of the square domain. The exact solution is

$$\psi(x, y) = 15(x^2 - x)(y^2 - y) - \sin 2\pi y \frac{\sinh 2\pi(x - 1)}{\sinh 2\pi}.$$

Let us solve the equation numerically using Fourier sine transform in the x direction. The dependent variable should have homogeneous boundary conditions at $x = 0$ and $x = 1$. Introducing a new variable $\phi(x, y)$ given by

$$\phi(x, y) = \psi(x, y) + (x - 1) \sin 2\pi y$$

results in a new Poisson equation

$$\frac{\partial^2 \phi}{\partial x^2} + \frac{\partial^2 \phi}{\partial y^2} = 30(x^2 - x) + 30(y^2 - y) - 4\pi^2(x - 1) \sin 2\pi y,$$

with $\phi(0, y) = \phi(1, y) = \phi(x, 0) = \phi(x, 1) = 0$. We now solve this equation for $M = N = 32$ ($\Delta_1 = \Delta_2 = 1/M$). For each j in (6.18), we use *Numerical Recipes*' `sinft` to obtain $\hat{Q}_{k,j}$, where $k = 1, 2, \ldots, (M - 1)$. For each k, we solve the tridiagonal system of equations (6.21). Finally, $\hat{\phi}_{k,j}$ is transformed to $\phi_{i,j}$ using `sinft` again. The solution of the original equation is then given by

$$\psi_{ij} = \phi_{ij} - (x_i - 1) \sin 2\pi y_j.$$

Both numerical and exact solutions are plotted in Figure 6.5. The two plots are indistinguishable; the maximum error is 0.001.

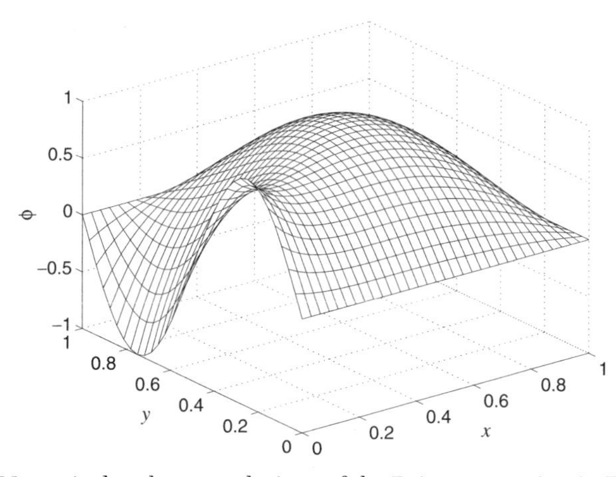

Figure 6.5 Numerical and exact solutions of the Poisson equation in Example 6.4.

6.2.2 Differentiation of a Periodic Function Using Fourier Spectral Method

The modified wavenumber approach discussed in Chapter 2 naturally points to the development of a highly accurate alternative to finite difference techniques: spectral numerical differentiation. Consider a periodic function $f(x)$ defined on N equally spaced grid points, $x_j = j\Delta$, and $j = 0, 1, 2, \ldots, N-1$. The spectral derivative of f is computed as follows. First, the discrete Fourier transform of f is computed as in (6.5)

$$\hat{f}_k = \frac{1}{N}\sum_{j=0}^{N-1} f_j e^{-ikx_j}$$

where,

$$k = \frac{2\pi}{L}n \quad n = -N/2, -N/2+2, \ldots, N/2-1.$$

Then, the Fourier transform of the derivative approximation is computed by multiplying the Fourier transform of f by ik

$$\widehat{Df}_k = ik\hat{f}_k \quad n = -N/2, -N/2+2, \ldots, N/2-1.$$

In practice, the Fourier coefficient of the derivative corresponding to the oddball wavenumber is set to zero, i.e., $\widehat{Df}_{-N/2} = 0$. This ensures that the derivative remains real in physical space (see Section 6.1.3), and it is only an issue when N is even.

Finally, the numerical derivative at a typical point j is obtained from inverse transformation

$$\left.\frac{\partial f}{\partial x}\right|_j = \sum_{k=-N/2}^{N/2-1} \widehat{Df}_k e^{ikx_j}.$$

It is easy to see that this procedure yields the *exact* derivative of the harmonic function $f(x) = e^{ikx}$ at the grid points if $|k| \le N/2 - 1$. In fact, the spectral derivative is more accurate than any finite difference scheme for periodic functions. The major cost involved is that of using the fast Fourier transform.

EXAMPLE 6.5 Differentiation Using the Fourier Spectral Method and Second-Order Central Difference Formula

(a) Consider the harmonic function $f(x) = \cos 3x$ defined on the discrete points $x_j = (2\pi/N)j$, where $j = 0, \ldots, N-1$. Its Fourier coefficients were calculated in Example 1(a). The Fourier coefficients of the derivative are given by $\widehat{Df}_k = ik\hat{f}_k$. They are therefore

$$\widehat{Df}_k = \begin{cases} -(3/2)i & \text{if } k = -3 \\ (3/2)i & \text{if } k = 3 \\ 0 & \text{otherwise.} \end{cases}$$

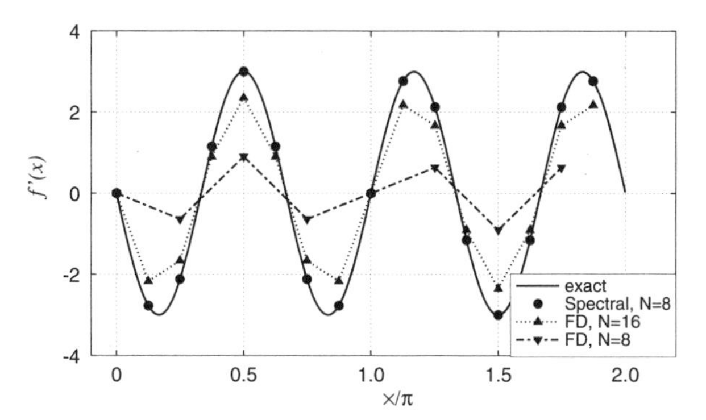

Figure 6.6 Numerical derivative of $\cos 3x$ in Example 6.5(a) using Fourier spectral method and second-order central finite difference formula (FD).

The corresponding inverse transform is $Df_j = -3\sin 3x_j$, which is the exact derivative of $f(x) = \cos 3x$ at the grid points. This exact answer is obtained as long as $N \geq 8$ (because $N/2 - 1 \geq 3$). For comparison, the second-order central difference formula (2.7) is also used to compute the derivative. Results are plotted in Figure 6.6 for $N = 8$ and 16 points. It is clear that the finite difference method requires many more points to give a result as accurate as the spectral method.

(b) Consider now the function $f(x) = 2\pi x - x^2$ defined on the same discrete set of points. We compute the Fourier coefficients of f_j using *Numerical Recipes*' `realft`, multiply $\hat{f}_k$ by ik, set the Fourier coefficient corresponding to $-N/2$ to zero, and finally inverse transform using `realft` to obtain the numerical derivative of f_j. Results are plotted in Figure 6.7 for $N = 16$. The finite difference derivative is exact since its truncation error for a quadratic is zero (see (2.7)). The spectral derivative is less accurate especially near the boundaries where the periodic continuation of $f(x)$ is discontinuous.

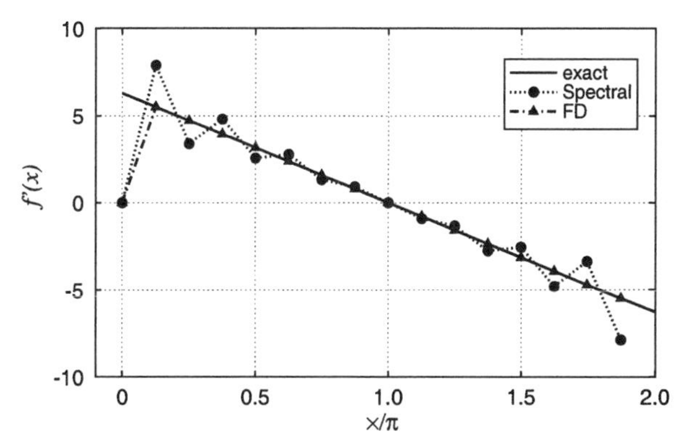

Figure 6.7 Numerical derivative of $2\pi x - x^2$ in Example 6.5(b) using Fourier spectral method and second-order finite differences (FD), with $N = 16$.

6.2.3 Numerical Solution of Linear, Constant Coefficient Differential Equations with Periodic Boundary Conditions

The Fourier differentiation technique is easily applied to the numerical solution of partial differential equations with periodic boundary conditions. Below we will present two examples, one for an elliptic equation, and another for an unsteady initial boundary value problem.

EXAMPLE 6.6 Poisson Equation

Consider the Poisson equation

$$\frac{\partial^2 P}{\partial x^2} + \frac{\partial^2 P}{\partial y^2} = Q(x, y) \tag{6.22}$$

in a periodic rectangle of length L_1 along the x axis and width L_2 along the y direction. Let us discretize the space with M uniformly spaced grid points in x and N grid points in y. The solution at each grid point is represented as

$$P_{l,j} = \sum_{n_1=-\frac{M}{2}}^{\frac{M}{2}-1} \sum_{n_2=-\frac{N}{2}}^{\frac{N}{2}-1} \hat{P}_{k_1,k_2} e^{ik_1 x_l} e^{ik_2 y_j}$$

$$l = 0, 1, 2, \ldots, M-1 \quad j = 0, 1, 2, \ldots, N-1, \tag{6.23}$$

where

$$x_l = lh_1, \quad h_1 = \frac{L_1}{M}, \quad y_j = jh_2, \quad h_2 = \frac{L_2}{N}, \quad k_1 = \frac{2\pi}{L_1} n_1, \quad k_2 = \frac{2\pi}{L_2} n_2.$$

Substituting (6.23) and the corresponding Fourier series representation for $Q_{l,j}$ into (6.22) and using the orthogonality of the Fourier exponentials, we obtain

$$-k_1^2 \hat{P}_{k_1,k_2} - k_2^2 \hat{P}_{k_1,k_2} = \hat{Q}_{k_1,k_2}, \tag{6.24}$$

which can be solved for $\hat{P}_{k_1,k_2}$ to yield

$$\hat{P}_{k_1,k_2} = -\frac{\hat{Q}_{k_1,k_2}}{k_1^2 + k_2^2}. \tag{6.25}$$

This is valid when k_1 and k_2 are not both equal to zero. The solution of the Poisson equation (6.22) with periodic boundary conditions is indeterminant to within an arbitrary constant. We can therefore set

$$\hat{P}_{0,0} = c,$$

where c is an arbitrary constant. Recall that $\hat{P}_{0,0}$ is simply the average of P over the domain (see 6.10). The inverse transform of $\hat{P}_{k_1,k_2}$ yields the desired solution $P_{l,j}$. Note that if we sum both sides of the Poisson equation with periodic boundary conditions over the domain, we get

$$\sum_{x_l} \sum_{y_j} Q(x_l, y_j) = 0.$$

Thus, the prescribed Q should satisfy this condition for the well posedness of the equation. An equivalent presentation of this condition is $\hat{Q}_{0,0} = 0$ (see (6.10)). This consistency condition can also be deduced from (6.24) by setting both wavenumbers equal to zero.

EXAMPLE 6.7 Initial Boundary Value Problem

(a) Consider the convection–diffusion equation

$$\frac{\partial u}{\partial t} + \frac{\partial u}{\partial x} = \nu \frac{\partial^2 u}{\partial x^2} + f(x, t) \tag{6.26}$$

in the domain $0 \leq x \leq L$, with periodic boundary conditions in x, and with initial condition $u(x, 0) = u_0(x)$. Since u is periodic in space, we will expand it in discrete Fourier series

$$u(x_j, t) = \sum_{n=-\frac{N}{2}}^{\frac{N}{2}-1} \hat{u}_k(t) e^{ikx_j}.$$

Substitution into (6.26) and using the orthogonality of the Fourier exponentials yields

$$\frac{d\hat{u}_k}{dt} = -(ik + \nu k^2)\hat{u}_k + \hat{f}_k(t).$$

This is an ordinary differential equation that can be solved for each $k = (2\pi/L)n$, with $n = 0, 1, 2, \ldots, N/2 - 1$, using a time advancement scheme. Here, we are assuming that u is real and therefore we need to carry only half the wavenumbers. The solution at any time t is obtained by inverse Fourier transformation of $\hat{u}_k(t)$.

(b) As a numerical example, we solve

$$\frac{\partial u}{\partial t} + \frac{\partial u}{\partial x} = 0.05 \frac{\partial^2 u}{\partial x^2},$$

on $0 \leq x \leq 1$ with

$$u(x, 0) = \begin{cases} 1 - 25(x - 0.2)^2 & \text{if } 0 \leq x < 0.4 \\ 0 & \text{otherwise.} \end{cases}$$

Let $N = 32$. We first use *Numerical Recipes*' `realft` to inverse transform $u(x_j, 0)$ and obtain $\hat{u}_k(0)$, $k = 2\pi n$, $n = 0, 1, 2, \ldots, N/2 - 1$. Next we advance in time the differential equation

$$\frac{d\hat{u}_k}{dt} = -(ik + 0.05k^2)\hat{u}_k$$

for each k using a fourth-order Runge–Kutta scheme. This equation is exactly the model equation we studied in Chapter 4, i.e., $y' = \lambda y$.

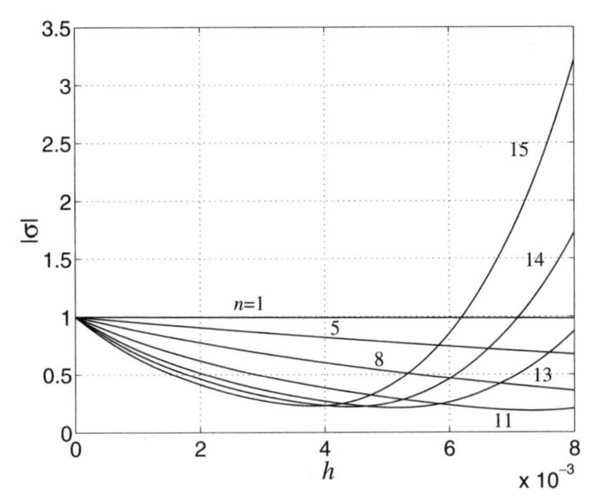

Figure 6.8 $|\sigma|$ versus h for $k = 2\pi n$, $n = 1, 5, 8, 11, 13, 14, 15$, in Example 6.7(b).

For stability, the time step h is chosen such that $\lambda h = -(ik + 0.05k^2)h$ falls inside the stability diagram of Figure 4.8. For fourth-order Runge–Kutta this means that

$$|\sigma| = \left|1 + \lambda h + \frac{\lambda^2 h^2}{2} + \frac{\lambda^3 h^3}{6} + \frac{\lambda^4 h^4}{24}\right| \leq 1.$$

If we plot $|\sigma|$ versus h for each k (see Figure 6.8), we find that as h increases, $|\sigma|$ becomes greater than 1 for the largest k value first, $k = 2\pi$ $(N/2 - 1)$. From the plot, the maximum value of h that can be used is 0.00620. In our calculation we used $h = 0.006$.

The solution is plotted in Figure 6.9 for $t = 0.25, 0.5$, and 0.75. The solution both propagates and diffuses in time, in accordance with the properties of the convective–diffusion equation.

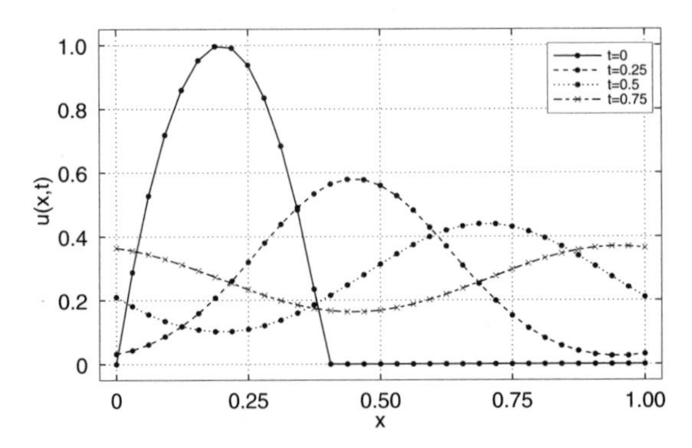

Figure 6.9 Numerical solution of the convective–diffusion equation in Example 6.7(b).

6.3 Matrix Operator for Fourier Spectral Numerical Differentiation

Up to this point we have described Fourier spectral numerical differentiation in terms of several steps: FFT of the function f, setting the oddball Fourier coefficient to zero, multiplying by ik, and inverse transforming back to the *physical space*. In some applications it is convenient or even necessary to have a compact representation of the spectral Fourier derivative operator in the physical space rather than the wave space. In this section we shall develop a physical space operator in the form of a matrix for numerical differentiation of a periodic discrete function and give an example of its application. This operator is, of course, completely equivalent to the familiar wave-space procedure.

Let u be a function defined on the grid

$$x_j = \frac{2\pi j}{N} \quad j = 0, 1, 2, \ldots, N-1.$$

Discrete Fourier transform of u is given by the following pair of equations:

$$\hat{u}_k = \frac{1}{N}\sum_{j=0}^{N-1} u(x_j)e^{-ikx_j} \tag{6.27}$$

and

$$u(x_j) = \sum_{k=-N/2}^{N/2-1} \hat{u}_k e^{ikx_j}.$$

Recall that the spectral derivative of u at the grid points is given by

$$(Du)_j = \sum_{k=-N/2+1}^{N/2-1} ik\hat{u}_k e^{ikx_j},$$

where the Fourier coefficient corresponds to the oddball wavenumber equal to zero (see Section 6.2.2). Substituting for $\hat{u}_k$ from (6.27) yields

$$(Du)_l = \frac{1}{N}\sum_{k=-N/2+1}^{N/2-1}\sum_{j=0}^{N-1} iku(x_j)e^{-ikx_j}e^{ikx_l} = \frac{1}{N}\sum_k\sum_j iku_j e^{\frac{2\pi ik}{N}(l-j)}$$

$$l = 0, 1, 2, \ldots, N-1.$$

Let

$$d_{lj} = \frac{1}{N}\sum_{k=-N/2+1}^{N/2-1} ike^{\frac{2\pi ik}{N}(l-j)} \quad l, j = 0, 1, 2, \ldots, N-1. \tag{6.28}$$

Then the derivative of u at each grid point is given by

$$(Du)_l = \sum_{j=0}^{N-1} d_{lj}u_j \quad l = 0, 1, 2, \ldots, N-1. \tag{6.29}$$

The right-hand side of this expression is in the form of multiplication of an $N \times N$ matrix D with elements d_{lj}, and the vector $\boldsymbol{u}$ with elements u_j. The matrix D is the physical space differentiation operator that we were after. We can simplify the expression for d_{lj} into a compact trigonometric expression without a summation. To evaluate the sum in (6.28), we first consider the geometric series

$$\begin{aligned}
S &= \sum_{k=-N/2+1}^{N/2-1} e^{ikx} = e^{i(-N/2+1)x} + e^{i(-N/2+2)x} + \cdots + e^{i(N/2-1)x} \\
&= e^{i(-N/2+1)x}\left[1 + e^{ix} + e^{2ix} + \cdots + e^{i(N-2)x}\right] \\
&= e^{i(-N/2+1)x}\frac{1 - e^{i(N-1)x}}{1 - e^{ix}} \\
&= \frac{e^{i(-N/2+1)x} - e^{i(N/2)x}}{1 - e^{ix}} \\
&= \frac{e^{i(-N/2+1/2)x} - e^{i(N/2-1/2)x}}{e^{-ix/2} - e^{ix/2}} \\
&= \frac{\sin\left(\frac{N-1}{2}x\right)}{\sin\frac{x}{2}}.
\end{aligned}$$

This expression can be differentiated to yield the desired sum

$$\frac{dS}{dx} = \sum_{k=-N/2+1}^{N/2-1} ike^{ikx} = \frac{\left(\frac{N-1}{2}\right)\cos\left(\frac{N-1}{2}x\right)\sin\frac{x}{2} - \frac{1}{2}\cos\frac{x}{2}\sin\left(\frac{N-1}{2}x\right)}{\left(\sin\frac{x}{2}\right)^2}.$$

The result can be further simplified by using the trigonometric identities

$$\begin{aligned}
\sin\left(\frac{Nx}{2} - \frac{x}{2}\right) &= \sin\frac{Nx}{2}\cos\frac{x}{2} - \cos\frac{Nx}{2}\sin\frac{x}{2} \\
\cos\left(\frac{Nx}{2} - \frac{x}{2}\right) &= \cos\frac{Nx}{2}\cos\frac{x}{2} + \sin\frac{Nx}{2}\sin\frac{x}{2},
\end{aligned}$$

and noting that in (6.28) we could make the following substitution:

$$x = \frac{2\pi}{N}(l - j).$$

After these substitutions and simplifications, we finally arrive at

$$\frac{dS}{dx} = \frac{N}{2}(-1)^{l-j}\cot\left[\frac{\pi(l-j)}{N}\right].$$

Thus, the matrix elements for Fourier spectral differentiation are

$$d_{lj} = \begin{cases} \frac{1}{2}(-1)^{l-j}\cot\left[\frac{\pi(l-j)}{N}\right] & \text{if } l \neq j \\ 0 & \text{if } l = j. \end{cases} \tag{6.30}$$

This result for the diagonal elements of the matrix is obtained directly from (6.28).

The problem of Fourier spectral differentiation has thus been converted to a matrix multiplication in physical space as in (6.29), and transformation to the wave space is not necessary. Recall from linear algebra (see Appendix) that multiplication of a full matrix and a vector requires $O(N^2)$ operations, which is more expensive than the $O(N \log_2 N)$ operations for the Fourier transform method. However, in some applications such as the numerical solution of differential equations with non-constant coefficients, having a derivative operator in the physical space is especially useful. Finite difference operators can also be written in matrix form, but they always lead to banded matrices. The fact that the Fourier spectral derivative operator is essentially a full matrix reflects the global or fully coupled nature of spectral differentiation: the derivative of a function at any grid point is dependent on the functional values at *all* the grid points.

EXAMPLE 6.8 Burgers Equation

We illustrate the use of the derivative matrix operator by solving the nonlinear Burgers equation

$$\frac{\partial u}{\partial t} + u\frac{\partial u}{\partial x} = \frac{\partial^2 u}{\partial x^2},$$

on $0 \leq x \leq 2\pi$, $0 < t \leq 0.6$ with $u(x, 0) = 10\sin(x)$ and periodic boundary conditions. Using explicit Euler for time advancement yields the discretized form of the equation as

$$\boldsymbol{u}^{n+1} = \boldsymbol{u}^n + h(D^2\boldsymbol{u}^n - UD\boldsymbol{u}^n),$$

where $\boldsymbol{u}^n$ is a column vector with elements u_j^n and $j = 0, \ldots, N-1$, D is a matrix whose elements are d_{lj} from (6.30), and U is a diagonal matrix formed from the elements of $\boldsymbol{u}^n$.

We estimate the time step h by performing stability analysis on the following linearized form of the Burgers equation:

$$\frac{\partial u}{\partial t} + u_{\max}\frac{\partial u}{\partial x} = \frac{\partial^2 u}{\partial x^2},$$

where $u_{\max}$ is the maximum absolute value of $u(x, t)$ over the given domain. We assume that the maximum value of $u(x, t)$ occurs at $t = 0$; that is, $u_{\max} = 10$. This assumption will be verified later by checking that the numerical solution of u does not exceed 10. Substituting the mode $\hat{u}_k(t)e^{ikx}$ for u, we have

$$\frac{d\hat{u}_k}{dt} = \lambda\hat{u}_k, \qquad \text{where} \quad \lambda = -k(k + iu_{\max}).$$

For stability of the explicit Euler method, the condition $|1 + \lambda h| \leq 1$ must be satisfied. This is equivalent to $(1 + h\lambda_R)^2 + (h\lambda_I)^2 \leq 1$ or

$$h \leq -2\frac{\lambda_R}{|\lambda|^2}.$$

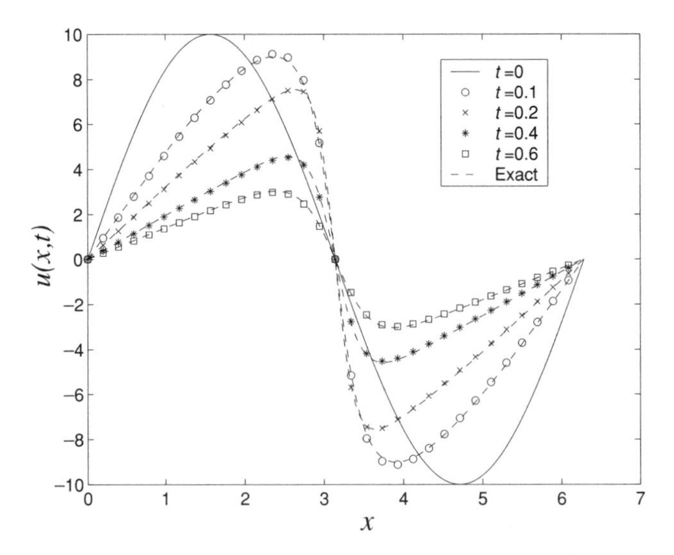

Figure 6.10 Numerical solution of the Burgers equation in Example 6.8.

Substituting $-k(k + iu_{\max})$ for λ gives

$$h \leq \frac{2}{k^2 + u_{\max}^2}.$$

The worst case scenario corresponds to the maximum value of $|k|$, i.e., $N/2$. For $N = 32$ and $u_{\max} = 10$, we obtain $h \leq 0.0056$. We use $h = 0.005$ in the present calculations. Solutions at $t = 0.1$, 0.2, 0.4, and 0.6 are shown in Figure 6.10.

The exact solution can be obtained from E. R. Benton and G. W. Platzman, "A table of solutions of the one dimensional Burgers equation," *Q. Appl. Math.* 30 (1972), p. 195–212, case 5. It is plotted in Figure 6.10 with dashed lines. The agreement is very good. In fact a similar agreement with the exact solution can be obtained with only $N = 16$.

The solution illustrates the main feature of the Burgers equation, which consists of a competition between convection and diffusion phenomena. The former causes the solution to steepen progressively with time, whereas the latter damps out high gradients. As a result, the solution first steepens and then slowly decays, as shown in Figure 6.10.

6.4 Discrete Chebyshev Transform and Applications

Discrete Fourier series are not appropriate for representation of non-periodic functions. When Fourier series are used for non-periodic functions, the convergence of the series with increasing number of terms is rather slow. In the remaining sections of this chapter we will develop the discrete calculus tools for non-periodic functions, using transform methods.

An arbitrary but smooth function can be represented efficiently in terms of a series of a class of *orthogonal polynomials* which are the eigenfunctions of the so-called *singular* Sturm–Liouville differential equations. Sines and cosines are examples of eigenfunctions of *non-singular* Sturm–Liouville problems. One of the advantages of using these polynomial expansions to approximate arbitrary functions is their superior resolution capabilities near boundaries. A rich body of theoretical work has established the reasons for excellent convergence properties of these series, which is outside the scope of this book. We will only use one class of these polynomials called Chebyshev polynomials.

An arbitrary smooth function, $u(x)$ defined in the domain $-1 \leq x \leq 1$ is approximated by a finite series of Chebyshev polynomials:

$$u(x) = \sum_{n=0}^{N} a_n T_n(x). \tag{6.31}$$

Chebyshev polynomials are solutions (eigenfunctions) of the differential equation

$$\frac{d}{dx}\left[\sqrt{1-x^2}\,\frac{dT_n}{dx}\right] + \frac{\lambda_n}{\sqrt{1-x^2}}T_n = 0,$$

where the eigenvalues $\lambda_n = n^2$. The first few Chebyshev polynomials are

$$T_0 = 1, \quad T_1 = x, \quad T_2 = 2x^2 - 1, \quad T_3 = 4x^3 - 3x, \ldots. \tag{6.32}$$

A key property of the Chebyshev polynomials is that they become simple cosines with the transformation of the independent variable $x = \cos\theta$, which maps $-1 \leq x \leq 1$ into $0 \leq \theta \leq \pi$. The transformation is

$$T_n(\cos\theta) = \cos n\theta. \tag{6.33}$$

This is the most attractive feature of Chebyshev polynomial expansions because the representation is reverted to cosine transforms, and in the discrete case one can take advantage of the FFT algorithm. Using a trigonometric identity, the following recursive relation for generating Chebyshev polynomials can be easily derived:

$$T_{n+1}(x) + T_{n-1}(x) = 2xT_n(x) \quad n \geq 1. \tag{6.34}$$

Other important properties of Chebyshev polynomials are

$$|T_n(x)| \leq 1 \qquad \text{in } -1 \leq x \leq 1, \quad \text{and}$$
$$T_n(\pm 1) = (\pm 1)^n.$$

To use Chebyshev polynomials for numerical analysis, the domain $-1 \leq x \leq 1$ is discretized using the "cosine" mesh:

$$x_j = \cos\frac{\pi j}{N} \quad j = N, N-1, \ldots, 1, 0. \tag{6.35}$$

It turns out that these grid points are the same as a particular set of Gauss quadrature points discussed in Chapter 2. If the problem is defined on a different domain than $-1 \le x \le 1$, the independent variable should be transformed to $-1 \le x \le 1$. For example, the domain $0 \le x < \infty$ can be mapped into $-1 \le \psi \le 1$ by the transformation:

$$x = \alpha \frac{1+\psi}{1-\psi}, \qquad \psi = \frac{x-\alpha}{x+\alpha},$$

where α is a constant parameter of the transformation.

As a direct consequence of the *discrete* orthogonality of cosine expansions, Chebyshev polynomials are discretely orthogonal under summation over $x_n = \cos(\pi n/N)$. That is

$$\sum_{n=0}^{N} \frac{1}{c_n} T_m(x_n) T_p(x_n) = \begin{cases} N & \text{if } m = p = 0, N \\ N/2 & \text{if } m = p \neq 0, N \\ 0 & \text{if } m \neq p, \end{cases}$$

where

$$c_n = \begin{cases} 2 & \text{if } n = 0, N \\ 1 & \text{otherwise.} \end{cases}$$

The discrete Chebyshev transform representation of a function u defined on a discrete set of points given by the cosine distribution in (6.35) is defined as

$$u_j = \sum_{n=0}^{N} a_n T_n(x_j) = \sum_{n=0}^{N} a_n \cos \frac{n\pi j}{N} \qquad j = 0, 1, 2, \ldots, N \tag{6.36}$$

where the coefficients are obtained using the orthogonality property by multiplying both sides of (6.36) by $(1/c_j)T_p(x_j)$ and summing over all j:

$$a_n = \frac{2}{c_n N} \sum_{j=0}^{N} \frac{1}{c_j} u_j T_n(x_j) = \frac{2}{c_n N} \sum_{j=0}^{N} \frac{1}{c_j} u_j \cos \frac{n\pi j}{N}$$

$$n = 0, 1, 2, \ldots, N. \tag{6.37}$$

Comparing (6.36) to (6.13), the Chebyshev coefficients for any function u in the domain $-1 \le x \le 1$ are exactly the coefficients of the cosine transform obtained using the values of u at the cosine mesh (6.35); i.e., $u_j = u[\cos(\pi j/N)]$.

EXAMPLE 6.9 Calculation of the Discrete Chebyshev Coefficients

We calculate the Chebyshev coefficients of x^4 and $4(x^2 - x^4)e^{-x/2}$ on $-1 \leq x \leq 1$ using *Numerical Recipes'* `cosft1`. As long as $N \geq 4$, the coefficients for x^4 are

$$\begin{cases} a_0 = 0.375, & a_2 = 0.5, \quad a_4 = 0.125 \\ a_n = 0 & \text{otherwise.} \end{cases}$$

We can validate this result as follows. Using (6.32) and (6.34), T_4 is given by

$$T_4 = 2xT_3 - T_2 = 2x(4x^3 - 3x) - T_2 = 8x^4 - 6x^2 - T_2.$$

Substituting $T_2 + T_0$ for $2x^2$ gives

$$x^4 = 0.375T_0 + 0.5T_2 + 0.125T_4,$$

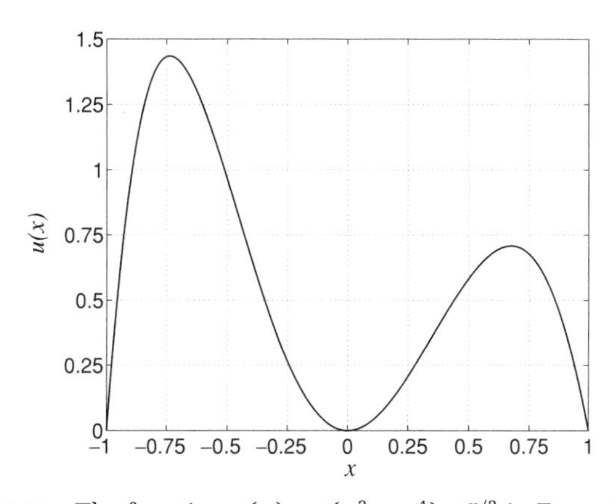

Figure 6.11 The function $u(x) = 4(x^2 - x^4)e^{-x/2}$ in Example 6.9.

which is in accordance with the coefficients obtained using `cosft1`. The function $u(x) = 4(x^2 - x^4)e^{-x/2}$ is plotted in Figure 6.11 and the magnitude of its Chebyshev coefficients for $N = 8$ are plotted in Figure 6.12. Strictly,

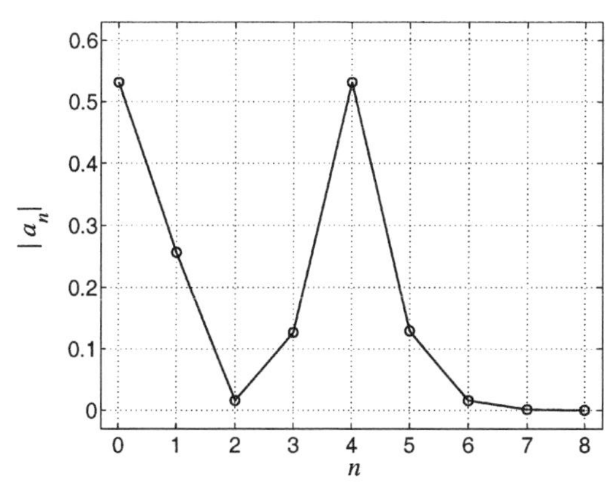

Figure 6.12 The magnitudes of the Chebyshev coefficients of $4(x^2 - x^4)e^{-x/2}$ in Example 6.9.

since u is not a polynomial it would have an infinite number of non-zero Chebyshev coefficients. However, the coefficients a_n are negligible for $n \geq 7$; i.e., only seven Chebyshev polynomials are needed to accurately represent $4(x^2 - x^4)e^{-x/2}$.

6.4.1 Numerical Differentiation Using Chebyshev Polynomials

The next step in the development of Chebyshev calculus is to derive a procedure for numerical differentiation of a function defined on the grid (6.35). Our objective is to obtain a recursive relationship between the coefficients of the Chebyshev transforms of a function and its derivative. In the case of Fourier expansion for a periodic function, this procedure was simply to multiply the Fourier transform of the function by ik. This is a bit more involved for Chebyshev representation, but not too difficult. Having the coefficients of the Chebyshev transform of the derivative, we obtain the derivative in the physical space on the grid (6.35) by inverse transformation.

We will first derive a useful identity relating the Chebyshev polynomials and their first derivatives. Recall from the definition of Chebyshev polynomials (6.35):

$$T_n(x) = \cos n\theta \qquad x = \cos\theta.$$

Differentiating this expression

$$\frac{dT_n}{dx} = \frac{d\cos n\theta}{d\theta}\frac{d\theta}{dx} = \frac{n\sin n\theta}{\sin\theta},$$

and using the trigonometric identity

$$2\sin\theta\cos n\theta = \sin(n+1)\theta - \sin(n-1)\theta,$$

we obtain the desired identity relating Chebyshev polynomials and their derivatives

$$2T_n(x) = \frac{1}{n+1}T'_{n+1} - \frac{1}{n-1}T'_{n-1} \quad n > 1. \tag{6.38}$$

Now consider the Chebyshev expansions of the function u and its derivative:

$$u(x) = \sum_{n=0}^{N} a_n T_n \tag{6.31}$$

$$u'(x) = \sum_{n=0}^{N-1} b_n T_n, \tag{6.39}$$

where a_n are the coefficients of u and b_n are the coefficients of its derivative. Note that since u is represented as a polynomial of degree N, its derivative can be a polynomial of degree at most $N-1$. Differentiating (6.31) and equating the result to (6.39) gives

$$\sum_{n=0}^{N-1} b_n T_n = \sum_{n=0}^{N} a_n T_n'.$$

Substituting for T_n using (6.38), we have

$$b_0 T_0 + b_1 T_1 + \sum_{n=2}^{N-1} b_n \frac{1}{2}\left[\frac{T_{n+1}'}{n+1} - \frac{T_{n-1}'}{n-1}\right] = \sum_{n=0}^{N} a_n T_n'. \tag{6.40}$$

Equating the coefficients of T_n', we finally obtain

$$\frac{b_{n-1}}{2n} - \frac{b_{n+1}}{2n} = a_n$$

or

$$b_{n-1} - b_{n+1} = 2na_n \quad n = 2, 3, \ldots, N-1, \tag{6.41}$$

where it is understood that $b_N = 0$ (see (6.39)). So far, we have $N-2$ equations for N unknowns. Equating the coefficients of T_N' on both sides of (6.40) yields

$$b_{N-1} = 2Na_N,$$

which is the same as we would obtain from (6.41), if we were to extend its range to N noting that $b_{N+1} = 0$. We still need one more equation. Noting that $T_1' = T_0$ and $T_2' = 4T_1$ from (6.40), we have

$$b_0 T_1' + \frac{1}{4} b_1 T_2' - \frac{b_2}{2} T_1' - \frac{1}{4} b_3 T_2' + \cdots = \sum_{n=0}^{N} a_n T_n'.$$

Equating the coefficients of T_1' from both sides gives

$$b_0 - \frac{1}{2} b_2 = a_1.$$

Hence, equation (6.41) can be generalized to yield all b_n as follows:

$$c_{n-1} b_{n-1} - b_{n+1} = 2na_n \quad n = 1, 2, \ldots, N \tag{6.42}$$

with $b_N = b_{N+1} = 0$.

In summary, to compute the derivative of a function u defined on the grid (6.35), one first computes its Chebyshev transform using (6.37), then the coefficients of its derivative are obtained from (6.42) by a straightforward marching from the highest coefficient to the lowest, and finally, the inverse transformation (6.36) is used to obtain u' at the grid points given by the cosine distribution.

A formal solution for the coefficients b_n in (6.42) can be written as

$$b_m = \frac{2}{c_m} \sum_{\substack{p=m+1 \\ p+m \text{ odd}}}^{N} p a_p. \tag{6.43}$$

The derivation of this equation is left as an exercise at the end of this chapter.

EXAMPLE 6.10 Calculation of Derivatives Using Discrete Chebyshev Transform

We want to calculate the derivatives of x^4 and $4(x^2 - x^4)e^{-x/2}$ defined on the cosine mesh inside the interval $-1 \leq x \leq 1$. We first calculate the coefficients b_n using (6.42) and the Chebyshev transform coefficients a_n already computed in Example 6.9. We then inverse transform b_n using `cosft1`, which is equivalent to (6.36), to obtain the derivative at the cosine mesh. For x^4 we obtain:

$$\begin{cases} b_1 = 3 & b_3 = 1 \\ b_n = 0 & \text{otherwise.} \end{cases}$$

This means that the derivative at the grid points is $3T_1(x_j) + T_3(x_j)$. From (6.32), this is equal to $4x_j^3$ which is the exact derivative of x^4 at the grid points.

The coefficients of the derivative of $4(x^2 - x^4)e^{-x/2}$ are computed and used to calculate the derivative, which is plotted in Figure 6.13 for $N = 5$. The results show good agreement with the exact derivative. For comparison, the derivative using second-order finite differences are also shown in Figure 6.13. In calculating the finite difference derivative, we use (2.7) for the interior grid points, (2.12) for the left boundary point, and

$$u'_j = \frac{3u_j - 4u_{j-1} + u_{j-2}}{2h}$$

at the right boundary point.

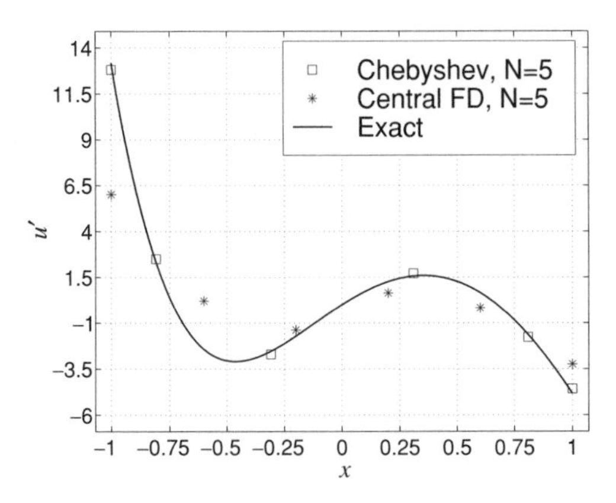

Figure 6.13 The derivative of $4(x^2 - x^4)e^{-x/2}$ using Chebyshev transform and central finite differences with $N = 5$ in Example 6.10.

6.4.2 Quadrature Using Chebyshev Polynomials

Equation (6.38) can also be used to derive a quadrature formula in a manner analogous to numerical differentiation. Integrating both sides of (6.38) leads to

$$\int T_n(x)\,dx = \begin{cases} T_1 + \alpha_0 & \text{if } n = 0 \\ \frac{1}{4}(T_0 + T_2) + \alpha_1 & \text{if } n = 1 \\ \frac{1}{2}\left[\frac{1}{n+1}T_{n+1} - \frac{1}{n-1}T_{n-1}\right] + \alpha_n, & \text{otherwise,} \end{cases}$$

where α_i are the integration constants. If u is represented by (6.31) and its definite integral $g(x) = \int_{-1}^{x} u(\xi)\,d\xi$ is represented by another Chebyshev expansion with coefficients d_n, then

$$\begin{aligned} g(x) = \int_{-1}^{x} u(\xi)\,d\xi &= \sum_{n=0}^{N+1} d_n T_n = \sum_{n=0}^{N} a_n \int T_n(x)\,dx \\ &= a_0 T_1 + \frac{a_1}{4}(T_0 + T_2) \\ &\quad + \sum_{n=2}^{N} \left\{ \frac{a_n}{2}\left[\frac{1}{n+1}T_{n+1} - \frac{1}{n-1}T_{n-1}\right] + \alpha_n \right\} + \alpha_0 + \alpha_1. \end{aligned}$$

Equating the coefficients of the same Chebyshev polynomials from both sides leads to the following recursive equation for the coefficients of the integral

$$d_n = \frac{1}{2n}(c_{n-1}a_{n-1} - a_{n+1}) \quad n = 1, 2, \ldots, N+1, \tag{6.44}$$

where it is understood that $a_{N+1} = a_{N+2} = 0$. All the integration constants and the coefficient of T_0 on the right-hand side can be combined into one integration constant that is equal to d_0. To obtain d_0, we note that $g(-1) = 0$, which leads to

$$\sum_{n=0}^{N+1} d_n(-1)^n = 0,$$

which can be solved for d_0 to yield

$$d_0 = d_1 - d_2 + d_3 - \cdots + (-1)^{N+2} d_{N+1}. \tag{6.45}$$

EXAMPLE 6.11 Calculation of Integrals Using Discrete Chebyshev Transform

We calculate the integrals

$$I_1 = \int_1^{\pi} \frac{\sin x}{2x^3}\,dx \quad \text{and} \quad I_2 = \int_1^{8} \frac{\log x}{x}\,dx$$

of Examples 3.1 and 3.4, respectively. The intervals of both integrals are transformed to $[-1, 1]$ by using the transformations or change of variables:

$$y = \frac{2x - (\pi + 1)}{\pi - 1} \quad \text{in} \quad I_1, \quad \text{and} \quad y = \frac{2x - 9}{7} \quad \text{in} \quad I_2.$$

The integrals then become

$$I_1 = \int_{-1}^{1} \frac{\pi - 1}{4} \frac{\sin[0.5(\pi - 1)y + 0.5(\pi + 1)]}{[0.5(\pi - 1)y + 0.5(\pi + 1)]^3} \, dy$$

and

$$I_2 = \int_{-1}^{1} \frac{7}{2} \frac{\log(3.5y + 4.5)}{3.5y + 4.5} \, dy.$$

These integrals are of the form $g(x) = \int_{-1}^{x} u(\xi) \, d\xi$. We first calculate a_n, the Chebyshev transform of the integrand $u(\xi)$, using `cosft1`. We then calculate d_n, the coefficients of its integral $g(x)$, from (6.44) and (6.45). Finally, we inverse transform d_n using `cosft1` to obtain $g(x)$ which can be evaluated at $x = 1$. In this case, we do not even need to inverse transform d_n to get $g(x)$ and then $g(1)$; $g(1)$ is simply equal to $\sum_{n=0}^{N+1} d_n$. The resulting % error in I_1 is 6.07×10^{-3} for $N = 8$ and 4.56×10^{-7} for $N = 16$, which is much lower than the error of any method in Example 3.1. The error ϵ in I_2 is 1.04×10^{-3} for $N = 8$ and 2.43×10^{-6} for $N = 16$. Comparing to Example 3.4, the Chebyshev quadrature performance is better than the performance of Simpson's rule but not as good as that of Gauss–Legendre quadrature.

6.4.3 Matrix Form of Chebyshev Collocation Derivative

As with Fourier spectral differentiation discussed in Section 6.3, it is sometimes desirable to have a physical space operator for numerical differentiation using Chebyshev polynomials. Consider the function $f(x)$ in the interval $-1 \leq x \leq 1$. We wish to compute the derivative of f on the set of collocation points $x_n = \cos \pi n/N$, with $n = 0, 1, 2, \ldots, N$. The discrete Chebyshev representation of f is given by

$$f(x) = \sum_{p=0}^{N} a_p T_p(x)$$

and

$$a_p = \frac{2}{N c_p} \sum_{n=0}^{N} \frac{1}{c_n} T_p(x_n) f_n \qquad c_n = \begin{cases} 2 & n = 0, N \\ 1 & \text{otherwise} \end{cases}$$

$$p = 0, 1, 2, 3, \ldots, N.$$

This expression can be written in matrix form for the vector of Chebyshev coefficients:

$$\boldsymbol{a} = \begin{bmatrix} a_0 \\ a_1 \\ \vdots \\ a_N \end{bmatrix} = \hat{T}\boldsymbol{f},$$

where

$$\hat{T} = \frac{2}{N} \begin{bmatrix} \frac{T_0(x_0)}{4} & \frac{T_0(x_1)}{2} & \dots & \frac{T_0(x_N)}{4} \\ \frac{T_1(x_0)}{2} & T_1(x_1) & \dots & \frac{T_1(x_N)}{2} \\ \vdots & \vdots & \vdots & \vdots \\ \frac{T_N(x_0)}{4} & \frac{T_N(x_1)}{2} & \dots & \frac{T_N(x_N)}{4} \end{bmatrix}.$$

Similarly the derivative of f is given by

$$f'(x_n) = \sum_{p=0}^{N} b_p T_p(x_n)$$

or

$$\boldsymbol{f}' = T\boldsymbol{b},$$

where

$$T = \begin{bmatrix} T_0(x_0) & T_1(x_0) & \cdots & T_N(x_0) \\ T_0(x_1) & T_1(x_1) & \cdots & T_N(x_1) \\ \vdots & \vdots & \vdots & \vdots \\ T_0(x_N) & T_1(x_N) & \cdots & T_N(x_N) \end{bmatrix}.$$

Recall that using (6.43), we can explicitly express the Chebyshev coefficients of f in terms of the Chebyshev coefficients of f':

$$b_p = \frac{2}{c_p} \sum_{\substack{n=p+1 \\ n+p \text{ odd}}}^{N} n a_n.$$

Again, in vector form this expression can be written as

$$\boldsymbol{b} = G\boldsymbol{a},$$

where

$$G_{pn} = \begin{cases} 0 & \text{if } p \geq n \text{ or } p+n \text{ even}, \\ \frac{2n}{c_p} & \text{otherwise.} \end{cases}$$

Thus, we have the following expression for f' at the collocation points:

$$\boldsymbol{f}' = TG\boldsymbol{a} = TG\hat{T}\boldsymbol{f} = D\boldsymbol{f},$$

where

$$D = TG\hat{T}. \tag{6.46}$$

The $(N+1) \times (N+1)$ matrix D is the desired physical space operator for Chebyshev spectral numerical differentiation. Multiplication of D by the vector consisting of the values of f on the grid results in an accurate representation of f' at the grid points. However, expression (6.46) for D is not very convenient because it is given formally in terms of the product of three matrices. It turns out that one can derive an explicit and compact expression for the elements of D using Lagrange polynomials as discussed in Chapter 1. This derivation is algebraically very tedious and is left as exercises for the motivated reader at the end of this chapter (Exercises 17 and 18); we simply state the result here. The elements of the $(N+1) \times (N+1)$ Chebyshev collocation derivative matrix D are

$$d_{jk} = \begin{cases} \frac{c_j(-1)^{j+k}}{c_k(x_j - x_k)} & j \neq k \\ \frac{-x_j}{2(1-x_j^2)} & j = k, \quad j \neq 0, N \\ \frac{2N^2+1}{6} & j = k = 0 \\ -\frac{2N^2+1}{6} & j = k = N, \end{cases} \tag{6.47}$$

where x_j are the locations of the grid points given by (6.35) and

$$c_j = \begin{cases} 2 & \text{if } j = 0, N \\ 1 & \text{otherwise.} \end{cases}$$

EXAMPLE 6.12 Calculation of Derivatives Using Chebyshev Derivative Matrix Operator

We use the Chebyshev derivative matrix operator to differentiate $u(x) = 4(x^2 - x^4)e^{-x/2}$ of Example 6.10. Let the vectors $\mathbf{x}$ and $\boldsymbol{u}$ represent the collocation points $x_n = \cos(\pi n/N)$, $n = 0, 1, 2, \ldots, N$, and the values of u at these points, respectively. For $N = 5$, $\mathbf{x}$ and $\boldsymbol{u}$ are

$$\mathbf{x} = \begin{bmatrix} 1.000 \\ 0.809 \\ 0.309 \\ -0.309 \\ -0.809 \\ -1.000 \end{bmatrix}, \qquad \boldsymbol{u} = \begin{bmatrix} 0 \\ 0.604 \\ 0.296 \\ 0.403 \\ 1.355 \\ 0 \end{bmatrix}.$$

The matrix operator D, whose elements are obtained from (6.47), is

$$D = \begin{bmatrix} 8.500 & -10.472 & 2.894 & -1.528 & 1.106 & -0.500 \\ 2.618 & -1.171 & -2.000 & 0.894 & -0.618 & 0.276 \\ -0.724 & 2.000 & -0.171 & -1.618 & 0.894 & -0.382 \\ 0.382 & -0.894 & 1.618 & 0.171 & -2.000 & 0.724 \\ -0.276 & 0.618 & -0.894 & 2.000 & 1.171 & -2.618 \\ 0.500 & -1.106 & 1.528 & -2.894 & 10.472 & -8.500 \end{bmatrix}.$$

We multiply D by $\boldsymbol{u}$ to obtain the derivative of $\boldsymbol{u}$ at the collocation points:

$$\boldsymbol{u}' = D\boldsymbol{u} = \begin{bmatrix} -4.581 \\ -1.776 \\ 1.717 \\ -2.703 \\ 2.502 \\ 12.813 \end{bmatrix}.$$

These values are exactly the ones obtained in Example 10 (see Figure 6.13).

EXAMPLE 6.13 Convection Equation with Non-constant Coefficients

We solve the equation

$$u_t + 2xu_x = 0 \qquad u(x,0) = \sin 2\pi x,$$

on the domain $-1 \le x \le 1$ using the matrix form of the Chebyshev collocation derivative to calculate the spatial derivatives. This is a one-dimensional wave equation with characteristics going out of the domain at both ends and thus there is no need for boundary conditions. Using the explicit Euler scheme for time advancement, the discretized form of the equation is

$$\boldsymbol{u}^{n+1} = \boldsymbol{u}^n + h(-2XD\boldsymbol{u}^n),$$

where $\boldsymbol{u}^n$ is a column vector with elements u_j^n, $j = 0, \ldots, N-1$; D is a matrix whose elements are d_{lj} from (6.47); and X is a diagonal matrix with x_j, $j = 0, \ldots, N-1$, on its diagonal.

For $N = 16$ and $h = 0.001$, solutions at $t = 0.3$ and 0.6 are shown in Figure 6.14. The agreement with the exact solution ($\sin(\pi x e^{-2t})$) is very good. Similar agreement can also be obtained with $N = 8$. From Figure 6.14, we see that the solution at the origin does not move. This is expected since the wave speed, x, is zero at $x = 0$. Also, the parts of the wave to the right and left of the origin propagated to the right and left, respectively. The wave shape is distorted since the speed of propagation is different from point to point.

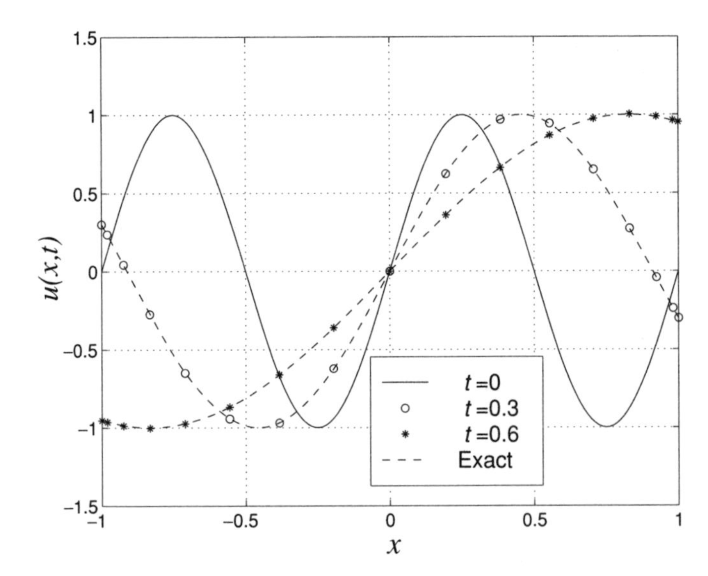

Figure 6.14 Numerical solution of the convection equation in Example 6.13

EXERCISES

1. Show that the Fourier coefficients of the discrete convolution sum

$$c_j = \sum_{n=0}^{N-1} f_n g_{j-n} = (f * g)_j$$

are given by

$$\hat{c}_k = N \hat{f}_k \hat{g}_k.$$

2. Consider the triple product defined by

$$B_{mn} = \sum_{j=0}^{N-1} u_j u_{j+m} u_{j+n}.$$

Show that the bi-spectrum, $\hat{B}_{k_1 k_2}$, the two-dimensional Fourier coefficients of B_{mn} are given by

$$\hat{B}_{k_1 k_2} = N \hat{u}_{k_1} \hat{u}_{k_2} \hat{u}_{-(k_1+k_2)}.$$

3. The discrete cosine series is defined by

$$f_j = \sum_{k=0}^{N} a_k \cos(kx_j) \quad j = 0, 1, 2, \ldots, N,$$

where $x_j = \pi j/N$. Prove that the coefficients of the series are given by

$$a_k = \frac{2}{N} \frac{1}{c_k} \sum_{j=0}^{N} \frac{1}{c_j} f_j \cos(kx_j) \quad k = 0, 1, 2, \ldots, N,$$

where

$$c_j = \begin{cases} 2 & j = 0, N \\ 1 & \text{otherwise.} \end{cases}$$

4. Given $H(x) = f(x)g(x)$, express the discrete cosine transform of H in terms of the discrete cosine transforms of f and g.
5. Use an FFT routine to compute the Fourier coefficients of

$$f(x) = \cos\frac{n\pi x}{L} \qquad 0 < x < L,$$

with $N = 8$, $L = 7$, and $n = 2, 3$. Use an FFT routine to compute the inverse transform of the coefficients to verify that the original data are recovered.
6. Compute the Fourier coefficients using FFT of

$$f(x) = \cos(2x) + \frac{1}{2}\cos(4x) + \frac{1}{6}\cos(12x) \qquad 0 \le x < 2\pi,$$

for $N = 8, 16, 32$, and 64.
7. Consider the function $f(x)$ defined as follows:

$$f(x) = \begin{cases} e^{-x} & \text{for } 0 \le x < L \\ 0 & \text{otherwise.} \end{cases}$$

Obtain the Fourier coefficients using FFT. Discuss the importance of L and N. In addition, compare the computational time of using the fast Fourier transform to the computational time of the brute-force ($O(N^2)$) Fourier transform. (Graph the computational time on a log–log plot.) To get good timing data, you may have to call the FFT routine several times for each value of N.
8. Differentiate the following functions using FFT and second-order finite differences. Show your results, including errors, graphically. Use $N = 16, 32$.
   (a) $f(x) = \sin 3x + 3\cos 6x \quad 0 \le x < 2\pi$.
   (b) $f(x) = 6x - x^2 \quad 0 \le x < 2\pi$.
9. Consider the ODE

$$f'' - f' - 2f = 2 + 6\sin 6x - 38\cos 6x,$$

defined on $0 \le x \le 2\pi$ with periodic boundary conditions and $f(0)=f(2\pi)=0$. Solve it using FFT and a second-order central finite difference scheme with $N = 16, 64$. Compare the results.
10. Discuss how to solve the following equation using the Fourier spectral method:

$$u_{xx} + (\sin x)u_x = -(\sin x + \sin 2x)e^{\cos x},$$

on $0 \le x \le 2\pi$ with periodic boundary conditions. Derive a set of algebraic equations for the Fourier coefficients. Be sure to carefully consider the boundary conditions and verify that the resulting matrix equation is non-singular.
11. Write a program that computes the Chebyshev transform of an arbitrary function, and test your program by transforming 1, x^3, and x^6. Use your program to compute and plot the Chebyshev expansion coefficients for

(a) $f(x) = xe^{-x/2}$.

(b) $f(x) = \begin{cases} +1 & -1 \le x \le 0 \\ -1 & 0 < x \le 1. \end{cases}$

Use $N = 4, 8$, and 16.

12. Write a program to calculate the derivative of an arbitrary function using the Chebyshev transform. Test your program by differentiating polynomials and use it to differentiate the functions in Exercise 11. Take $N = 4, 8, 16, 32$ and compare to the exact answers.

13. Use mathematical induction to show that

$$b_m = \frac{2}{c_m} \sum_{\substack{p=m+1 \\ p+m \text{ odd}}}^{N} p a_p,$$

where a_p are the Chebyshev coefficients of some function $f(x)$ and b_m are the Chebyshev coefficients of $f'(x)$.

14. Use the Chebyshev transform program in Exercise 11 to calculate the integral of an arbitrary function. Test your program by integrating polynomials and use it to integrate the functions in Exercise 11. Take $N = 4, 8, 16, 32$ and compare to the exact values.

15. Use the matrix form of the Chebyshev collocation derivative to differentiate $f(x) = x^5$ for $-1 \le x \le 1$. Compare to the exact answer.

16. Solve the convection equation

$$u_t + 2u_x = 0,$$

for $u(x, t)$ on the domain $-1 \le x \le 1$ subject to the boundary and initial conditions

$$u(-1, t) = \sin \pi t \qquad u(x, 0) = 0.$$

The exact solution is

$$u = \begin{cases} 0 & x \ge -1 + 2t \\ \sin \pi (t - \frac{x+1}{2}) & -1 \le x \le -1 + 2t. \end{cases}$$

Use the discrete Chebyshev transform and second-order finite difference methods. Plot the solution at several t. Plot the rms of the error at $t = 7/8$ versus N. Compare the accuracy of the two methods.

17. Show that the interior $N - 1$ Chebyshev grid points given by (6.35) are the zeros of T'_N which is a polynomial of degree $N - 1$.

18. In this exercise we will go through the key steps leading to expression (6.47) for the elements of the Chebyshev derivative matrix. We will begin by using the results from Exercise 10 of Chapter 1. Let $\phi_{N+1}(x)$ be a polynomial of

degree $N+1$:

$$\phi_{N+1}(x) = \prod_{l=0}^{N}(x - x_l).$$

Show that the matrix elements obtained in Exercise 10 of Chapter 1 can be recast in the following form:

$$d_{jk} = \frac{\phi'_{N+1}(x_j)}{\phi'_{N+1}(x_k)(x_j - x_k)} \quad j \neq k. \tag{1}$$

If $x_0 = -1$, $x_N = 1$, and the remaining x_j are the zeros of the polynomial $Q_{N-1}(x)$, then

$$\phi_{N+1}(x) = (1 - x^2)Q_{N-1}. \tag{2}$$

Show that

$$d_{jk} = \frac{\left(1 - x_j^2\right)Q'_{N-1}(x_j)}{\left(1 - x_k^2\right)Q'_{N-1}(x_k)(x_j - x_k)} \quad j \neq k \text{ and } j, k \neq 0, N \tag{3}$$

For $j = k$, again referring to Exercise 10 of Chapter 1, we want to evaluate

$$d_{jj} = \sum_{\substack{l=0 \\ l \neq j}}^{N} \frac{1}{x_j - x_l}.$$

Let $\phi_{N+1}(x) = (x - x_j)g(x)$, and let x_k ($k = 0, 1, 2, \ldots, N$ except for $k = j$) be the zeros of g. Show that

$$\frac{g'(x_j)}{g(x_j)} = \frac{\phi''_{N+1}(x_j)}{2\phi'_{N+1}(x_j)}, \tag{4}$$

and hence

$$d_{jj} = \frac{\phi''_{N+1}(x_j)}{2\phi'_{N+1}(x_j)}.$$

For Chebyshev polynomials, $x_0 = -1$, $x_N = 1$, and the remaining x_j are the zeros of T'_N (see Exercise 17). Using the fact that Q_{N-1} in (2) is simply equal to T'_N, you should now be able to derive the matrix elements given in (6.47), i.e.,

$$d_{jk} = \begin{cases} \frac{c_j(-1)^{j+k}}{c_k(x_j - x_k)} & j \neq k \\ \frac{-x_j}{2\left(1 - x_j^2\right)} & j = k, \quad j \neq 0, N \\ \frac{2N^2+1}{6} & j = k = 0 \\ -\frac{2N^2+1}{6} & j = k = N. \end{cases}$$

FURTHER READING

Bracewell, R. N. *The Fourier Transform and Its Applications*, Second Edition. McGraw-Hill, 1986.

Canuto, C., Hussaini, M. Y., Quarteroni, A., and Zang, T. A. *Spectral Methods in Fluid Dynamics*. Springer-Verlag, 1988.

Dahlquist, G., and Björck, Å. *Numerical Methods*. Prentice-Hall, 1974, Chapter 9.

Gottlieb, D., and Orszag, Steven A. *Numerical Analysis of Spectral Methods: Theory and Applications*. Society for Industrial and Applied Mathematics (SIAM), 1977.

Hockney, R. W., and Eastwood, J. W. *Computer Simulation Using Particles*: IOP (Inst. of Physics) Publishing Ltd. 1988, reprinted 1994.

Press, W. H., Teukolsky, S. A., Vetterling, W. T., and Flannery, B. P. *Numerical Recipes: The Art of Scientific Computing*, Second Edition. Cambridge University Press, 1992, Chapters 12 and 13.

Snyder, M. A. *Chebyshev Methods in Numerical Approximation*. Prentice-Hall, 1966, Chapters 1, 2, and 3.

APPENDIX

A Review of Linear Algebra

This appendix contains a brief review of concepts in linear algebra used in the main body of the text. Although numerical linear algebra lies at the foundation of numerical analysis, it should be the subject of a separate course. The intent of this appendix is to provide a convenient brush up on elementary linear algebra for the reader who has been previously exposed to this very important subject.

A.1 Vectors, Matrices and Elementary Operations

A vector is an ordered array of numbers or algebraic variables. In column form the vector $\boldsymbol{c}$ is represented as

$$\boldsymbol{c} = \begin{bmatrix} c_1 \\ c_2 \\ c_3 \\ \vdots \\ c_n \end{bmatrix}.$$

The vector $\boldsymbol{c}$ has n elements and has dimension n. The row vector $\boldsymbol{c}$ is simply written as

$$\boldsymbol{c} = [c_1, c_2, c_3, \ldots, c_n].$$

The inner product (or scalar product) of two n-dimensional real vectors $\boldsymbol{u}$ and $\boldsymbol{v}$ is defined as

$$(\boldsymbol{u}, \boldsymbol{v}) = u_1 v_1 + u_2 v_2 + \cdots + u_n v_n = \sum_{i=1}^{n} u_i v_i.$$

The length or the norm of the real vector $\boldsymbol{u}$ is the square root of its inner product with itself:

$$||\boldsymbol{u}|| = \sqrt{(\boldsymbol{u}, \boldsymbol{u})} = \sqrt{u_1^2 + u_2^2 + \cdots + u_n^2}.$$

The vectors $\boldsymbol{u}^1, \boldsymbol{u}^2, \boldsymbol{u}^3, \ldots, \boldsymbol{u}^n$ are said to be linearly independent when it is impossible to represent any one of them as a linear combination of the others. In other words if $a_1\boldsymbol{u}^1 + a_2\boldsymbol{u}^2 + \cdots + a_n\boldsymbol{u}^n = 0$ and the a_i are constant, then all a_i must be zero.

A matrix is a doubly ordered array of elements. An $m \times n$ matrix A has m rows and n columns and is written as

$$A = \begin{bmatrix} a_{11} & a_{12} & a_{13} & \ldots & a_{1n} \\ a_{21} & a_{22} & a_{23} & \ldots & a_{2n} \\ \vdots & & & & \\ \vdots & & & & \\ a_{m1} & a_{m2} & a_{m3} & \ldots & a_{mn} \end{bmatrix}.$$

The matrix elements are a_{ij}, where $i = 1, 2, \ldots, m$, and $j = 1, 2, \ldots, n$. If $\boldsymbol{v}$ is a vector of dimension n, the product of the $m \times n$ matrix A and the vector $\boldsymbol{v}$ is a vector $\boldsymbol{u}$ of dimension m, which in vector form is written as

$$A\boldsymbol{v} = \boldsymbol{u}.$$

The elements of $\boldsymbol{u}$ are

$$u_i = \sum_{j=1}^{n} a_{ij} v_j \quad i = 1, 2, \ldots, m. \tag{A.1}$$

Vector $\boldsymbol{u}$ can also be written as a linear combination of the columns of A, which are designated by $\boldsymbol{a}^i$:

$$\boldsymbol{u} = v_1\boldsymbol{a}^1 + v_2\boldsymbol{a}^2 + \cdots + v_n\boldsymbol{a}^n.$$

The product of A and an $n \times l$ matrix B is the $m \times l$ matrix C with elements computed as follows:

$$c_{ij} = \sum_{k=1}^{n} a_{ik} b_{kj} \quad i = 1, 2, \ldots, m \quad j = 1, 2, \ldots, l.$$

In general matrix multiplication is not commutative. That is, if A and B are $n \times n$ square matrices, in general, $AB \neq BA$.

The *identity matrix*, denoted by I, is a square matrix whose diagonal elements are 1 and off-diagonal elements are zero. The *inverse* of a square matrix A, denoted by A^{-1}, is defined such that $AA^{-1} = I$. A *singular matrix* does not have an inverse. The *transpose* of a matrix A, denoted by A^T, is obtained by exchanging the rows with columns of A. That is, the elements of A^T are $a_{ij}^T = a_{ji}$. A *symmetric matrix* A is equal to its transpose, i.e., $A = A^T$. If $A = -A^T$ then A is called *anti-symmetric* or *skew-symmetric*.

Application of most numerical discretization operators to differential equations leads to *banded matrices*. These matrices have non-zero elements in a narrow band around the diagonal of the matrix, and the rest of the elements are zero. A *tridiagonal matrix* has a non-zero diagonal and two adjacent sub- and super-diagonals:

$$A = \begin{bmatrix} b_1 & c_1 & & & \\ a_2 & b_2 & c_2 & & \\ & \ddots & \ddots & \ddots & \\ & & a_{n-1} & b_{n-1} & c_{n-1} \\ & & & b_n & c_n \end{bmatrix}.$$

The notation $B[a_i, b_i, c_i]$ is sometimes used to denote a tridiagonal matrix. Similarly a pentadiagonal matrix can be denoted by

$$B[a_i, b_i, c_i, d_i, e_i],$$

where c_i are the diagonal elements. An $n \times n$ tridiagonal matrix can be stored using $3n$ words as compared to n^2 for a *full matrix*. As will be pointed out later, working with tridiagonal and other banded matrices is particularly cost effective.

The *determinant* of a 2×2 matrix is defined as

$$\det \begin{bmatrix} a_{11} & a_{12} \\ a_{21} & a_{22} \end{bmatrix} = a_{11}a_{22} - a_{12}a_{21}.$$

For an $n \times n$ matrix the determinant can be calculated by the so-called row or column expansions:

$$\det A = \sum_{j=1}^{n} (-1)^{i+j} a_{ij} M_{ij} \quad \text{for any } i,$$

or

$$\det A = \sum_{i=1}^{n} (-1)^{i+j} a_{ij} M_{ij} \quad \text{for any } j.$$

M_{ij} is called the co-factor of the element a_{ij}, it is the determinant of the matrix formed from A by eliminating the row and column to which a_{ij} belongs. This formula is recursive; it is used on the subsequent smaller and smaller matrices until only 2×2 matrices remain for which their determinant is already given.

In modern linear algebra, the determinant is primarily used in analysis and to test for the singularity of a square matrix. A square matrix is singular if its determinant is zero. It can be shown that the determinant of the product of two matrices is equal to the product of their determinants. That is, if A and B are square $n \times n$ matrices, then

$$\det(AB) = \det(A)\det(B).$$

Thus, if any one of the two matrices is singular, their product is also singular.

A.2 System of Linear Algebraic Equations

A system of n algebraic equations in n unknowns is written as

$$A\boldsymbol{x} = \boldsymbol{b},$$

where A is an $n \times n$ matrix, $\boldsymbol{x}$ is the n dimensional vector of unknowns and $\boldsymbol{b}$ is the n dimensional right-hand side vector. If A is non-singular, the formal solution of this system is $\boldsymbol{x} = A^{-1}\boldsymbol{b}$. However, the formal solution which involves computation of the inverse is almost never used in computer solution of a system of algebraic equations. Direct numerical solution using computers is performed by the process of *Gauss elimination* which is a series of row operations. First, a set of row operations, called the *forward sweep*, uses each diagonal element as *pivot* to eliminate the elements of the matrix below the diagonal. Next, *backward substitution* is used to obtain the solution vector, starting from x_n to x_1.

The matrix A has a unique decomposition into upper and lower triangular matrices

$$A = LU,$$

where L is lower and U is upper triangular matrices. The elements of L and U are readily obtained from Gauss elimination. If the system of equations $A\boldsymbol{x} = \boldsymbol{b}$ is to be solved several times with different right-hand sides, then it would be cost effective to store L and U matrices and use them for each right-hand side. This is because the Gauss elimination process for triangular matrices does not require the forward sweep operations and therefore is much less expensive (see Section A.3). Suppose A is decomposed, then the system of equations is written as

$$LU\boldsymbol{x} = \boldsymbol{b}.$$

Let $\boldsymbol{y} = U\boldsymbol{x}$, then the equations are solved by first solving for $\boldsymbol{y}$

$$L\boldsymbol{y} = \boldsymbol{b}$$

and then for $\boldsymbol{x}$ using

$$U\boldsymbol{x} = \boldsymbol{y}.$$

Both of these steps involve only triangular matrices, which are significantly cheaper to solve.

A.2.1 Effects of Round-off Error

Round-off error is always present in computer arithmetic and can be particularly damaging when solving a system of algebraic equations. There are usually two types of problems related to the round-off error: one is related to the algorithm, i.e., the way Gauss elimination is performed, and the other is due to the matrix itself. In the elimination process, one ensures that each diagonal element (pivot) has a larger magnitude than all the elements below it, which are eliminated in the forward sweep. This is accomplished by scaling the elements of each row (including the right-hand side vector) so that the largest element in each row is equal to 1, and by *row exchanges*. This process is called *pivoting* and is used in most software packages.

Ill-conditioning refers to the situation where the matrix in the system of algebraic equations is nearly singular. In this case, slight errors in the right-hand side vector (which could be due to round-off error or experimental error) can amplify significantly. In other words, a small perturbation to the right-hand side vector can result in a significant change in the solution vector. The condition number of the matrix is a good indicator of its "condition." The condition number of A is defined as

$$\gamma(A) = \|A\| \cdot \|A^{-1}\|,$$

where $\|A\|$ is the norm of A. There are many ways to define the norm of a matrix. One example is the square root of the sum of the squares of its elements. If A and B are square matrices of the same size, $\boldsymbol{x}$ is a vector, and α is a real number, the norm must satisfy these properties: $\|A\| \geq 0$, $\|\alpha A\| = |\alpha| \, \|A\|$, $\|A + B\| \leq \|A\| + \|B\|$, $\|AB\| \leq \|A\| \cdot \|B\|$, and $\|A\boldsymbol{x}\| \leq \|A\| \cdot \|\boldsymbol{x}\|$. The matrix norm associated with the vector norm defined earlier is denoted by $\|A\|_2$ and is equal to the square root of the maximum eigenvalue of $A^T A$.

The condition number is essentially the amplification factor of errors in the right-hand side. Generally, round-off errors can cause problems if the condition number is greater than the relative accuracy of computer arithmetic. For example, if the relative accuracy of the computer is in the fifth decimal place, then the condition number of 10^5 or larger is cause for alarm.

A.3 Operations Counts

One of the important considerations in numerical linear algebra is the number of arithmetic operations required to perform a task. It is easy to count the number of multiplications, additions (or subtractions), and divisions for any algorithm. In the following we assume that all matrices are $n \times n$ and vectors have dimension n.

It can be easily verified from (A.1) that multiplication of a matrix and a vector requires n^2 multiplications and $n(n-1)$ additions. For large n we would say that *multiplication of a matrix and a vector* requires $O(n^2)$ of both additions and multiplications. Similarly, *multiplication of two matrices* requires $O(n^3)$ of both additions and multiplications.

With a bit more work it can be shown that solving a system of algebraic equations by Gauss elimination requires

- $\frac{1}{3}n^3 + \frac{1}{2}n^2 - \frac{5}{6}n$ of both additions and multiplications, and
- $\frac{1}{2}n(n+1)$ divisions.

Thus, for large n the Gauss elimination process for an arbitrary full matrix requires $O(n^3)$ operations which is substantial. However, most of the work is done in the forward sweep. Of the total number of operations, the forward elimination process alone requires $\frac{1}{3}(n^3 - n)$ additions and multiplications and $\frac{1}{2}n(n-1)$ divisions. Thus the backward elimination requires only $O(n^2)$ operations which is an insignificant part of the overall work for large n. This is why once a matrix is decomposed into LU, the solution process for different right-hand side vectors is rather inexpensive. There is also a significant reduction in the number of operations when solving systems with banded matrices. In Gauss elimination one simply takes advantage of the presence of zero elements and does not operate on them. For example, solving a system with a *tridiagonal* matrix requires $3(n-1)$ additions and multiplications and $2n-1$ divisions. This is a tremendous improvement over a general matrix.

A.4 Eigenvalues and Eigenvectors

If A is an $n \times n$ matrix, the eigenvalues of A are defined to be those numbers λ for which the equation

$$Ax = \lambda x$$

has a non-trivial solution x. The vector x is called an eigenvector belonging to the eigenvalue λ. The eigenvalues are the solutions of the *characteristic equation*,

$$\det(A - \lambda I) = 0.$$

The characteristic equation is a polynomial of degree n. The eigenvalues can be complex and may not be distinct. The characteristic equation can be used to show that the determinant of A is the product of its eigenvalues

$$\det(A) = \lambda_1 \lambda_2 \lambda_3 \ldots \lambda_n.$$

From this result it can be seen that a singular matrix must have at least one zero eigenvalue. In practice one does not actually use the characteristic equation to find the eigenvalues; the so-called QR algorithm is usually the method of choice and is the basis for computer programs available in numerical analysis libraries for computing eigenvalues and eigenvectors. If an $n \times n$ matrix has n distinct eigenvalues, $\lambda_1, \lambda_2, \ldots, \lambda_n$, then it has n linearly independent eigenvectors, $x^1, x^2, \ldots, x^n$. Moreover, the eigenvector x^j belonging to eigenvalue λ_j is unique apart from a non-zero constant multiplier. However, an $n \times n$ matrix may have n linearly independent eigenvectors, even if it does not have n distinct eigenvalues.

Two matrices A and B are called *similar* if there exists a non-singular matrix T such that:

$$T^{-1}AT = B.$$

Similar matrices have the same eigenvalues with the same multiplicities. If A has n linearly independent eigenvectors, then it is similar to a diagonal matrix, which according to the similarity rule, has the eigenvalues of A on the diagonal:

$$S^{-1}AS = \Lambda = \begin{bmatrix} \lambda_1 & 0 & 0 & \ldots & 0 \\ 0 & \lambda_2 & 0 & \ldots & 0 \\ \vdots & & \ddots & & \\ \vdots & & & \ddots & \\ 0 & 0 & 0 & \ldots & \lambda_n \end{bmatrix}.$$

The columns of the matrix S are the eigenvectors of A. This similarity transformation is an important result that is often used in numerical analysis. This transformation is also sometimes referred to as the diagonalization of A, which can be used to uncouple linear systems of coupled differential or difference equations. From the similarity transformation we can obtain an expression for powers of matrix A

$$A^k = S\Lambda^k S^{-1}.$$

Thus, if the moduli of the eigenvalues of A are less than 1, then

$$\lim_{k\to\infty} A^k \to 0.$$

This important result is true for all matrices, whether they are diagonalizable or not, as long as the magnitudes of the eigenvalues are less than 1.

Symmetric matrices arise frequently in numerical analysis and in modeling physical systems and have special properties which are often exploited. The *eigenvalues of a symmetric matrix are real* and eigenvectors belonging to different eigenvalues are orthogonal. An $n \times n$ symmetric matrix has n independent eigenvectors and therefore is always diagonalizable. If the eigenvectors are properly normalized so that they become orthonormal, then S^{-1} in the similarity transformation is simply S^T.

Index